Managing the Development
of New Products

Managing the Development of New Products

Achieving Speed and Quality Simultaneously Through Multifunctional Teamwork

Milton D. Rosenau, Jr., CMC

John J. Moran, CMC

JOHN WILEY & SONS, INC.

New York • Chichester • Weinheim • Brisbane • Singapore • Toronto

Library of Congress Cataloging-in-Publication Data:

Rosenau, Milton D., 1931–
    Managing the development of new products: achieving speed and quality simultaneously through multifunctional teamwork / Milton D. Rosenau, Jr., John J. Moran.
        p.   cm.
    Includes bibliographical references and index.
    ISBN 0-471-29183-8
    1. Industrial development projects—Management.   2. New products—Management.   3. Work groups.   I. Title.
HD69.P75R668   1993
658.5'75—dc20   93-14968

Printed in the United States of America

10 9 8 7 6 5

Contents

Preface

WHO THIS BOOK IS FOR

This book is for the new product development manager. It is also for members of the multifunctional team charged with the task of defining and bringing a high-quality new product or service concept to market quickly.

In many ways, new product development team managers are like orchestra conductors. They probably have extensive experience in a functional area (most likely marketing, engineering, or operations), but the challenge as a project manager is different. The essence of the new assignment is to understand and integrate the efforts of different functional groups. The multifunctional project team must get the new product out on time to meet the needs of a fickle, competitive marketplace. Quite a plateful!

This book is intended to help you if you have been trained in a professional skill and now find yourself thrust into a position of managing a new product development project in which you've had little or no direct experience. Now you must not only demonstrate your professional skill, but must take responsibility for schedule and budget, using physical and human resources over which you may have little or no real control, frequently including other companies that supply components of your product. You may be a marketing specialist, product line manager, engineer, scientist, or manufacturing expert asked to take responsibility for the new product effort. No matter what your previous experience, if you're now responsible for getting something done by a specified date within a limited budget, this book will help you grasp and master key practical skills for successful new product development management.

Although this book is aimed primarily at the recently appointed manager of a new product development project, other professionals considering such a position will also find much of value. People who are already new product development project managers with only limited project management experience will gain a better understanding or learn new techniques to improve their effectiveness. Functional managers who must support new product development efforts will gain important insights.

Most of the discussion and examples are specific to small and medium-sized projects, rather than to very large ones. However, large projects are normally managed as a collection of smaller, integrated ones; so the book has relevance regardless of project size.

THIS BOOK'S APPROACH

This book helps the reader identify and integrate four important considerations that are critical to successful new products:

Speed to market
Quality
Profitability
The multifunctional team

The first two considerations determine whether your new product will be attractive—or at least acceptable—to prospective purchasers. The third consideration is obviously your company's *raison d'être* and must be a key objective for you and your multifunctional project team. No company can remain in business, however attractive or necessary its products and services are, unless it continues to earn an adequate financial return. Thus, you must integrate and manage the multifunctional project team to produce new products that satisfy both an external market's needs and your company's goals.

Software is rapidly becoming indispensable in facilitating the new product development project manager's job. At various stages, the new product development project manager must be coach, businessperson, engineer, and diplomat. An understanding of the available software and how it can help organize and structure this complicated process is essential.

We have identified four general categories of software that can help the new product development project manager:

Project management software: software products that help to clarify work breakdown structures, schedules, costs, and resource requirements
Strategic and business planning software: software packages that help organize the assessment of the business viability of the product (Generally based on decision analysis or expert systems, they help tie together and document market, competitive and technical assumptions, and trade-offs.)
Financial modeling software: generally spreadsheet-based software that puts the market, cost, and availability data together to generate a measure of the financial

attractiveness of the new product (The financial model is usually tied closely with output from the business planning and other cost models.)

Team-based software: a general category of relatively new software products aimed at facilitating the multifunctional team decision-making process (Specific packages range from the general, such as idea generation and consensus measurement, to the specific, such as Quality Function Deployment [QFD].)

We outline and define the nature of these four important development elements and improve the understanding of them and how they fit together. Most important, we help the new product development manager and the "orchestra" meet their challenging task.

No treatment can or should trivialize a subject as complex as new product project management, but this book is specifically designed to make it as simple as possible. Simple project management tools can help you get high-quality new products and services to market fast by promoting coordinated cooperation in a multifunctional project team. We provide a step-by-step approach that we have found effective in our combined more than sixty years of industrial experience. Since the founding of our respective consulting practices, we have also had the benefit of teaching thousands of seminar students, working adults of all ages from a wide variety of industries. By taking you through the new product development project management process in eighteen steps (Chapters 2–19), the book aims to equip you with detailed tools that you can immediately apply to your first (or next) new product project and that will help you overcome the pitfalls that typically bedevil the new product development project manager. These tools will help you deal with *both* technical and mechanical situations (such as the work breakdown structure, the schedule, and the budget) *and* conceptual or people issues (such as the concerns of top management and motivation).

USEFUL AND UNIQUE FEATURES
OF THIS BOOK

There are many books on single, specialized aspects of project management (for instance, PERT and CPM) and several books on project management in specialized applications (for instance, the construction or aerospace industries), but this book is intended specifically for the people who manage or are involved in new product and service development projects of any description.

For simplicity, the book covers the subject matter chronologically, from a project's beginning to its end. Each short chapter is devoted to a single topic and can be absorbed in one to two hours. Thus, the entire book can be mastered in a single month, making it uniquely useful for the working adult. We also provide references to other books and articles that can amplify our discussion if you want additional material.

The book will be useful for almost any type of new product project and has many practical examples. In addition, we use some very basic illustrations to assure that a few critical issues are easily grasped, regardless of the reader's specific technical specialization.

There are six appendices, containing abbreviations, a glossary of terms, helpful checklists for project managers, and more detailed and extensive discussions of three important and helpful tools: financial analysis, project management software, and quality function deployment. Employing these three software tools will not make you a successful new product development project manager, and utilizing these is no guarantee of being a good project manager. Nevertheless, you can be a more effective new product development project manager if you use such software when it will help you. The examples are intended to illustrate some ways to use this software to your advantage. Although our illustrations come from specific versions of current mass-market project management and other software, our goal is not to demonstrate these products per se or teach how they may be used. Rather, we want to explain a few of the key useful aspects of these types of software.

HOW THIS BOOK IS ORGANIZED

Our step-by-step approach divides project management into five general managerial activities (or processes) and emphasizes the importance of satisfying the three constraints of the product's total performance specifications, schedule for its time to market, and development budget (which affects its profitability). The chapter sequence is a good match for the chronological concerns during a typical new product project. Your situation may differ, in which case you may wish to read the chapters out of sequence.

The first managerial activity is defining the project. Under this topic, we introduce the Triple Constraint and discuss strategic business and quality considerations in starting successful new product projects.

The second activity is planning. Chapters covering why and how to plan, the work breakdown structure, scheduling tools and time estimating, network diagrams (PERT, CPM, precedence, and others), planning the budget, the impact of limited resources, and contingency and risk are included in our discussion of planning.

Leading is the third managerial activity; and as part of this topic, we discuss how to organize a project, organizing the multifunctional project team and varied support team members, the role of the project manager, and some practical tips for project managers.

The fourth activity is monitoring. Monitoring tools, reviews of projects, handling changes, and solving the inevitable problems are covered under this subject.

Under completing, the fifth managerial activity, we tell you how to complete a new product development project and accomplish the final wrap-up.

Milton D. Rosenau, Jr., CMC
Rosenau Consulting Company
Houston, Texas

John J. Moran, CMC
J.J. Moran & Associates
Westlake Village, California

Acknowledgments

We wish to express our thanks to our clients and numerous colleagues in industry who provided so many helpful suggestions and illustrations. This book also benefits from questions asked and ideas provided by thousands of executive education and seminar students, so we appreciate the support of the various university, association, and other sponsors of these seminars.

A few of the particular project management software illustrations (of SuperProject 2.0 and Project Workbench 3.1) that we use in Appendix F were kindly created by Bill Stewart of the Stewart-Gordan Consulting Group, Atlanta. One illustration (of Project Workbench 3.1) was provided by Phil Wolf of Applied Business Technology Corporation, New York. Our goal is not to demonstrate specific products per se or teach how they may be used. Rather, the software illustrations demonstrate general capabilities and principles. Nevertheless, we consider them to be good packages in their respective market niches. However, all microcomputer software and hardware is changing and improving rapidly, and you should not interpret our illustrative use of any specific software products as an endorsement of or recommendation for their use in your own situation.

Anamaria Hildebrand transcribed large portions of our manuscript, and we appreciate her support. Finally, our publisher obtained anonymous reviews of an early version of the draft manuscript, and the book has benefited greatly from these reviewers' comments. Whatever errors or confusion may still remain are obviously our responsibility.

It is also a pleasure to acknowledge the fine support provided by our publisher, Robert L. Argentieri, and his staff. This book would not have come into being without his encouragement and help.

PART 1

Overview

CHAPTER 1

Quality, Speed, and Multifunctional Teamwork

In this chapter, we first review the critical importance of new products for companies and connect that to the modern quality movement. Next we discuss the imperative to bring new products to market quickly and the essential role of the multifunctional project team in making that possible. We also introduce the need for a new product financial analysis that shows a prospective profit sufficiently great to justify the inherent risk of new product development.

THE IMPORTANCE OF NEW PRODUCTS

New products are the cornerstone of the long-term survival and prosperity of most firms. However, getting a high-quality product to the right customer in a timely manner is a task that receives less than satisfactory marks in most organizations. Why? Why is the success rate—however defined—for all new products so low? Is it a question of competition, fickle markets, or organization effectiveness? Although the first two items certainly exist, they are outside our control. So we turn our attention to the third. How can companies improve the management of this critical activity? How can they change the effectiveness and the culture of their organizations to develop more profitable products?

For many firms, a change in culture, a new model—a "paradigm shift"—in the way they approach new product development is in order. The types of changes necessary include the following:

From	*To*
Manage results	Manage the process
Failure = punishment	Failure = learning
Short term	Long term
Individual	Team
Functional orientation	Team orientation
Technical performance	Customer satisfaction

These changes in corporate culture are presented as a *direction,* not an objective. Management's key responsibility is to strike the appropriate balance between the two extremes in the list. Company values and culture will take time to change. However, marketplace efficiency is often a big incentive.

The changes we are describing go far beyond new product development; they address the very core of corporate and eventually national competitiveness. However, there is an important opportunity in new product development that should not be overlooked.

New product development projects are in many ways minibusinesses facing the same multifunctional coordination challenges as the parent firm. But new product development is also a *project*-oriented activity with specific time-related tasks and objectives. This temporary nature of the new product development process positions it well as a catalyst for change within the firm. The temporary nature of the project organization makes change easier for it than for the rest of the organization.

Functional organizations have a very different orientation than project-oriented organizations (teams). The functional organization has a vested interest in its own longevity, "professional" status, and growth, as well as the level of value provided for its customers. These motivations are very often at the core of the top-heavy, bureaucratic organization. This is precisely the value system or culture that must change if the corporation is to survive. The action and task nature of new product development and project management "cultures" can well become a model and a catalyst for affecting that change.

We must be willing to rethink the way we develop new products and have the courage to make the changes necessary.

New product development offers the opportunity of becoming a catalyst for corporate change.

QUALITY

The common thread and direction for many of the changes for new product development are contained in the quality movement. Quality has been a major concern of most companies for a long time. Until recently, however, quality as applied to new products had a fairly narrow scope. The emphasis was on quality assurance and quality control, and the major keeper of product quality was the manufacturing department. The focus was on controlling product defects and maintaining design tolerances. To a large extent, we "inspected in" quality to the new product at the end of the production line. In the past several years, there has been a dramatic change; a major new element—the customer—has entered the quality equation.

The new quality direction has a relatively simple message. The measure of quality is customer satisfaction. W. Edwards Deming, widely recognized as the "father" of the quality movement, makes the point that "good quality" and "quality control" have no meaning unless related to a customer's needs.

Although the customer has always been an integral part of the new product development process, the recent emphasis represents a significant opportunity for those firms having the vision to grasp its significance. The umbrella for this new emphasis on quality is often called "Total Quality Management" (TQM). Although the interpretations of TQM often vary, the results of a successful TQM approach are unmistakable:

Quality has been expanded from inspection to include recognizing customer satisfaction.

- Engineering shifts its emphasis from technology to include the "voice" of the customer.
- Marketing recognizes that, in addition to closing the sale, creating a dialogue with the customer is important.

TQM represents the elements of a new paradigm for the firm, and its new product development efforts in particular. It represents a subtle shift—from the management of quality to the quality of management. The elements of TQM contain the basic approaches for companies to use in developing new products, indeed, in the basic running of their business.

TQM can be best understood by breaking it down into its components:

Focus	*System*
Strategic	Vision
	Management by walking around (MBWA)
	Participative management
Cultural	Multifunctional teams
	Statistical thinking
	Consensus decisions
Technical	Statistical process control (SPC)
	Concurrent engineering
	Quality function deployment (QFD)

The focus of TQM affects three areas of the new product development process. At a strategic level, the thinking and actions of the new product development project manager take on a new perspective both externally regarding the customer and internally managing available resources. At the implementation of the new product development process, through people, TQM provides for a new culture to take root. It both respects the individual and harnesses the power of the multifunction team. The third area can best be thought of as a "toolbox." These tools can be used to facilitate TQM and the new product development process. Other examples of the tools you may wish to use include consulting services, focus groups, market surveys, brainstorming, benchmarking, Pareto charts, and Taguchi methods.

TQM affects the new product development process in four basic areas:

1. The overall management of the new product development strategic, cultural, and team direction and emphasis.
2. The acquisition and integration of external market and competitive information in planning the new product
3. The management of the internal design, manufacturing, and support of the new product
4. The facilitation of the new product development process, particularly as it applies to improving interdisciplinary communication

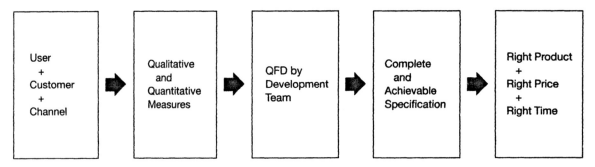

FIGURE 1-1.   One approach to assure that the new product's specification satisfies the market requirement.

The last area represents a significant opportunity for the new product development project manager. Applying TQM tools such as QFD can have a positive impact on both the time to market and customer acceptance of the new product. QFD is a powerful graphic planning and documentation tool, used by the multifunctional new product development team to deploy the customer's voice throughout the new product development process. We further describe QFD in Chapter 4, and Appendix E provides a more detailed description.

Figure 1-1 shows a flow of corporate activity that can assure the new product has appropriate quality to satisfy the totality of market requirements. Qualitative and quantitative market research (for example, focus groups and conjoint analysis, respectively) assures that QFD starts with the voice of the customer, leading to a complete product specification that the multifunctional project team believes is achievable.

Total Quality Management (TQM) represents a new way of new product development that can increase customer satisfaction, reduce time to market, and lower development expense.

THE MARKETING-QUALITY INTERFACE

Perhaps the most important interface to be managed in the new product development process is the translation of market needs into product specifications. It is useful to think of this interface in terms of obtaining the definition of "quality" that exists in the customer's mind. The purpose of market research then becomes to define what quality means to the target market segment. A 1987 article defines nine broad "quality dimensions":

Dimension of Quality	*Example*
1. Performance	Product speed
2. Conformance	Meets industry standard
3. Features	Secondary standards ("bells and whistles")
4. Reliability	Consistency over time (failure rate)
5. Serviceability	Resolution of problems (easy to repair)
6. Durability	Stands up over time (useful life)
7. Esthetics	Appearance (look)
8. Response	Timeliness (professionalism)
9. Reputation	Past performance (intangibles)

It also describes how the customer's voice is used to define each by providing specific examples.[1]

The concept here is an important one, for it not only lets the customer lead in developing the product specification, but also focuses initial thinking on a market segment—application dimension—*before* getting to the more specific product—feature dimension. The linkages developed with this approach can be invaluable in developing the initial customer requirements for the QFD matrix.

QUALITY AND RESULTS

There have been several efforts to look at the effect of quality on firms.[2] A May 1991 Government Accounting Office study compiled data using criteria from the Malcolm Baldrige Award to investigate the impact of TQM on corporate performance. Responses were collected from twenty firms that had received Baldrige site visits in 1988 and 1989. Respondents reported their perceptions of the impact of their quality efforts in a number of areas, including the following:

- Employee relations (attendance, turnover)
- Operating indicators (reliability, product lead time)
- Customer satisfaction (customer complaints, retention)
- Financial performance (market share, sales per employee)

Although the results were positive across all areas, none of the companies reported immediate benefits. Allowing sufficient time for results to be achieved was an important consideration.

Another major study was recently completed by Ernst and Young and the American Quality Foundation. The project was undertaken to document how quality affects the way businesses operate. The project looked at ten selected quality processes in more than five hundred businesses in Canada, Germany, Japan, and the United States. Top line findings important to the new product development professional include these:

Despite the emphasis on customer satisfaction, few companies have fully implemented programs to include the customer in the strategic planning, engineering, and product development efforts. Japan and Germany translate customer expectations into the design of new products much more frequently than either U.S. or Canadian firms.

Japanese firms had the highest rate of employee participation in quality teams. Canadian, German, and U.S. firms expected to increase employee involvement in teams.

Half the Japanese companies use process simplification and cycle reduction processes more than 90 percent of the time. Fewer than 25 percent of the businesses in the United States, Canada, or Germany were able to make this claim.

This last finding highlights a major difference between Japanese and U.S. management approaches to new product development. American management is *results* oriented, whereas Japanese management is *process* oriented. The Japanese word for this process approach is *kaizen,* meaning improvement.

By emphasizing results to the exclusion of process, a lot can be lost. This is particularly true in the new product development process, where success or failure is frequently in the hands of the marketplace or competitive reactions.

The ability to improve the process continually is actually an important enabler of improved results. The successful firm will blend the big change of innovation and the continuous improvement of process development into the new product area.

MANAGING FAST NEW PRODUCT DEVELOPMENT PROJECTS

The product life cycle (PLC) is getting shorter for many—perhaps all—products and services. More rapid and pervasive communication assures that every new product's availability is known worldwide quickly, sometimes instantaneously. Swiftly changing technology and totally new technological breakthroughs (for instance, the radio frequency–excited light bulb and high-temperature superconductivity) allow developers easily to improve product features and benefits.[3]

Many firms are responding to this challenge by shortening the time it takes to get a high-quality new product to market. Despite a stellar record of previous successes, Compaq Computer Corporation "ousted the company's president and chief executive" because he was blamed for taking too long to get new products to market.[4] The chairman of 3M recently stated that they intend to halve product introduction time, largely by strengthening multifunctional teamwork.[5] Figure 1-2 summarizes some North American improvements in time to market. Although the data supporting this

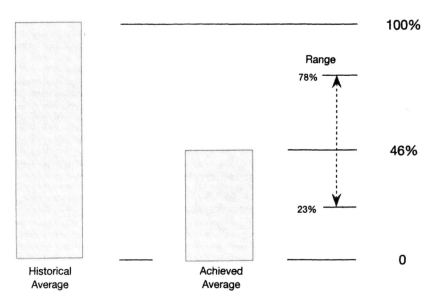

FIGURE 1-2.   Reported time-to-market improvement for twenty-seven North American new product developments.

illustration are somewhat informal, because they depend on the imperfect memory of those involved rather than unambiguous measurement, the overall trend is both clear and dramatic.

The techniques employed in the cases illustrated in Figure 1-2 are summarized in Table 1-1 (see following page). In all but one of the cases for which information is available, teamwork in some form is cited as a factor that allowed the faster development time. As we stress throughout this book, multifunctional teamwork involving *as a minimum* personnel from the triad of (1) marketing, (2) research, development, and engineering (R, D, and E), and (3) manufacturing (or the equivalent functions for services) is critical.

Figure 1-3 indicates the relationships between the four critical activities that may follow the new product idea. These activities are feasibility, new product development, maintenance, and a post mortem. This "roadmap" provides an overview of the total new product development process, and we link many of our ideas to it.

An idea, of itself, is not a new product, nor does it inherently justify a new product development effort. A feasibility activity—without any commitment to product development (fast or otherwise)—should be undertaken if there are any unknowns about the market, implementation technology, or production and delivery process. Major *unknowns* requiring feasibility efforts might include the desire to enter a new market, embody a totally new and unfamiliar technology in the product, or employ a production process or machine with which the company has no experience. The objective of this feasibility activity is to remove the unknowns (to permit new product development) or record the remaining unknowns in the form of improved knowledge for future use. The latter is

A successful new product idea may involve four activities: feasibility, development, maintenance, and a post mortem.

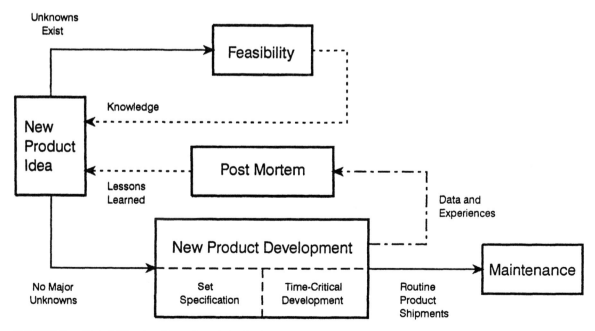

FIGURE 1-3. Pictorial flowchart illustrating relationship between critical new product development activities.

TABLE 1-1. Detailed Information About Reported Time-to-Market Improvements for Data Shown in Figure 1-2

	Previous	Achieved	Achieved as Percent of Previous	Source of Information
Norstar modular telephone	60 months	14 months	23	18
Warner Electric clutch brake	36 months	36 weeks	23	2
Honeywell thermostat	48 months	12 months	25	1
Harris Broadcast radio products	36 months	9 months	25	11
IBM personal computer	48 months	13 months	27	12
Allen-Bradley electrical contactors	6 years	2 years	33	1
CalComp vector plotter	42 months	14 months	33	24
Allen-Bradley printed circuit board	22 weeks	8 weeks	36	20
Motorola pocket pager	48 months	18 months	38	5 & 13
Cincinnati Milacron machining center	30 months	12 months	40	4
H-P computer printer	54 months	22 months	41	5 & 9
Ingersoll Rand air-powered grinder	36 months	15 months	42	8
P&G Pampers Phases	27 months	12 months	44	21
Intel chip	108 weeks	48 weeks	44	14
Codex communication products	145 weeks	67 weeks	46	6
3M digital color printer	6 years	3 years	50	16
Xerox copier	6 years	3 years	50	3
Navistar truck	5 years	30 months	50	5
NCR checkout counter terminals	44 months	22 months	50	10
AT&T phone	2 years	1 year	50	5
Whistler radar detector	21 months	11 months	52	15
General Electric jet engine	7 years	4 years	57	19
Chrysler 1992 Viper	5 years	3 years	60	17
IBM (Lexington, KY) varied products	28 months	18 months	64	7
GM 1991 Buick Park Avenue	60 months	40 months	67	12
GM 1992 Buick LeSabre	49 months	34 months	69	23
Chrysler LH cars	54 months	42 months	78	22
Total (weeks)	5,299	2,444	46 overall average	

Information Sources (chronological order)

1. *Wall Street Journal,* 23 February 1988, p. 1ff.
2. *Industry Week,* 1 August 1988, pp. 73–74.
3. *Wall Street Journal,* 21 November 1988, p. A14.
4. *R&D* magazine, January 1989, pp. 15–16.
5. *Fortune,* 13 February 1989, pp. 53–59.
6. *Electronic Business,* 29 May 1989, p. 25.
7. Talk by Kailash C. Joshi at CalTech Quality/Productivity Forum, 12 June 1989.
8. *New York Times,* 25 March 1990, sec. 3, pp. 1ff.
9. Talk by Don Palmer at PDMA-WEST Conference, 23 February 1990.
10. *Business Week,* 30 April 1990, pp. 110ff.
11. Talk by Frank A. Svet at PDMA 14th Anual International Conference, 2 November 1990.
12. P. G. Smith and D. G. Reinertsen, *Developing Products in Half the Time.* New York: Van Nostrand, 1991, p. 2.
13. K. B. Clark and T. Fujimoto, *Product Development Performance.* Boston: Harvard Business School Press, 1991, p. 350.
14. *Electronic Business,* 17 June 1991, p. 29.
15. *Electronic Business,* 17 June 1991, p. 66.
16. *Business Week,* 16 September 1991, pp. 59ff.
17. *Business Week,* 4 November 1991, pp. 36ff.

TABLE 1-1
(continued)

Techniques mentioned (in source) to shorten time to market								
MFT	CE	TW	OPDP	IF/SW	QFD	RPDP	A	Other
MFT	CE	TW	OPDP					Monthly project reviews and electronic data interchange
MFT	CE			IF/SW				
MFT			OPDP					
MFT		TW	OPDP				A	Early models and firm specifications
		TW		IF/SW				Built-in flexibility
MFT	CE							
MFT	CE	TW	OPDP					
								Information technology
MFT	CE	TW						Monthly prototypes
MFT				IF/SW				Modules
					QFD			Artificial intelligence and computer networks
MFT	CE	TW						Product family
								Chairman's pressure
(no specific techniques cited in source)								
MFT			OPDP					
MFT								
(no specific techniques cited in source)								
	CE	TW						
	CE							
MFT								
MFT			OPDP		QFD			
MFT		TW				RPDP		Work breakdown structure and program control board
MFT	CE	TW		IF/SW				Innovative materials
MFT					QFD	RPDP	A	Simulation
MFT						RPDP		
MFT	CE	TW						DFMA
MFT	CE	TW						Early prototype and no finger-pointing

18. S. R. Rosenthal, *Effective Product Design and Development.* Homewood, IL: Business One Irwin, 1992, pp. 187-211.
19. S. R. Rosenthal, *Effective Product Design and Development.* Homewood, IL: Business One Irwin, 1992, pp. 235-262.
20. S. R. Rosenthal, *Effective Product Design and Development.* Homewood, IL: Business One Irwin, 1992, p. 273.
21. *Business Week,* 3 February 1992, pp. 54-56.
22. *Los Angeles Times,* 26 April 1992, pp. D1ff.
23. *Industry Week,* 18 May 1992, pp. 46-53.
24. Talk by William T. Cloake at PDMA-WEST Conference, 10 July 1992.

Abbreviations for Techniques

MFT	Multifunctional teams
CE	Concurrent (or simultaneous) engineering
TW	Teamwork with suppliers
OPDP	Orderly product development process
IF/SW	Isolated facility or skunk works
QFD	Quality function deployment
RPDP	Reuse of previously developed parts
A	Automation
DFMA	Design for manufacturing and assembly

especially valuable if new technology overcomes a current limitation or competitive conditions change. A new product development activity can commence if the multifunctional development team believes there are no major unknowns. (Unfortunately, of course, major *problems*, as distinct from recognized unknowns, may still turn up sometime in development, frequently causing consternation if not a delay in product introduction.)

Feasibility activities may be carried out solely by a single function. An investigation of a totally new (unknown) market might initially involve only marketing personnel. Research to determine the range of useful properties of a new material or behavior of a new biological substance might be performed only by scientific personnel. The testing of new automated factory equipment to determine its suitability might be carried out only by manufacturing engineers.

MULTIFUNCTIONAL TEAMWORK

Unlike the feasibility activity, the new product development activity *must* be carried out by multifunctional teams, as emphasized in Figure 1-4. Furthermore, the development activity is composed of two (or more) discrete phases. The first is to optimize the trade-off between product specifications and the date of first shipment. The objective is to establish the specification for what is to be developed. Given defined resources, a

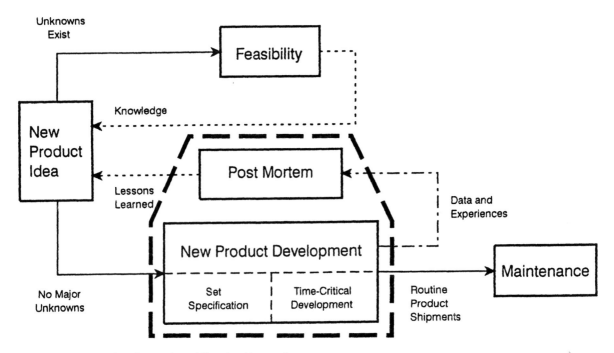

FIGURE 1-4.   Activities that require multifunctional teamwork.

product with difficult specifications requires a longer time to market, whereas a product with easier specifications can be developed more quickly, as illustrated in Figure 1-5. More effective resources (internal or external) usually can shorten the time to market for a given specification; less effective resources will lengthen it.

Figure 1-3 depicts the time allocation within the new product development activity for "set specification" and "time-critical development" as equal, but either can be much longer than the other. If the specification is relatively obvious but the development is difficult, the latter would be much longer than the former. The reverse could be true if setting the specification involved complex market research and many iterations of laboratory investigation and model making, followed by the decision quickly to introduce a simple product intended to be the initial member of an entire family of new products.

Because the multifunctional team—and its company—must be concerned with the effort's profitability, a financial analysis is required. Figure 1-6 illustrates a typical discounted cash flow financial analysis (the details of which are explained in Appendix D) that can tie together the following:

The degree to which the product specification matches the true market need, which is the main determinant of sales

The actual and perceived quality of the product that comes out of the new product development process when shipments commence, which also affects sales

The time of market entry, which partially determines both sales and development expense

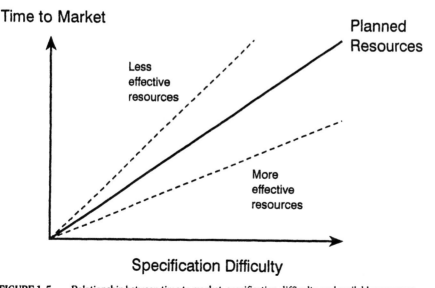

FIGURE 1-5.   Relationship between time to market, specification difficulty, and available resources.

YEAR =	1	2	3	4	5
1 Company Sales		1.00	2.00	4.00	1.00
2 Manufacturing Cost		.40	.80	1.20	.30
3 Development Expense	1.00	.50			
4 Operating Expense		.20	.40	.60	.10
5 Capital Expense	.50				
6 Depreciation	.10	.10	.10	.10	.00
7 Gross Profit	.00	.60	1.20	2.80	.70
8 Before Tax Income	-1.10	-.20	.70	2.10	.60
9 Income Tax	-.55	-.10	.35	1.05	.30
10 Net Income	-.55	-.10	.35	1.05	.30
11 Operating Cash Flow	-.45	.00	.45	1.15	.30
12 Working Cash	.00	-.30	-.30	-.60	.90
13 Total Cash Flow	-.95	-.30	.15	.55	1.50
14 Cumulative Cash Flow	-.95	-1.25	-1.10	-.55	.95

Internal Rate of Return (%) =   18.35

Discounted Cash Flows:

NPV @ 10%	.34
NPV @ 15%	.12
NPV @ 20%	-.05
NPV @ 25%	-.20
NPV @ 30%	-.32
NPV @ 35%	-.41
NPV @ 40%	-.50

FIGURE 1-6.   A typical discounted cash flow financial analysis.

The multifunctional project team's design and manufacturing trade-offs, which determine the product's cost and affect quality

Other corporate costs (marketing, sales, and general expenses, for example), which can impact sales

The extent to which the new product requires incremental capital investments to facilitate production or distribution

Some companies make the mistake of doing a discounted cash flow analysis very early in the new product development process. This is premature because so little is known about critical details (such as those listed in the previous paragraph). For example, it requires about a dozen years to bring a new drug to market, and perhaps five thousand are tested to obtain one approval by the Federal Drug Administration.[6] Although the $200+ million to fund such development work clearly must be justified, the information on which to do a meaningful financial analysis obviously is not available at the very early stages.

If the introduction date is critical (for instance, to exhibit at a major trade show), the corporation may have to commit more resources than originally contemplated, or the development team may have to relax the specification. Figure 1-7 highlights this interaction. Such trade-offs also may impact the financial return.

The trade-off between time to market and specification difficulty depends on available resources and may affect the effort's profitability.

Once the specification is set, the time-critical part of the development activity can start. This may be subdivided into further phases such as product design, process design, pilot production, and so forth, as we discuss in Chapter 3. Concurrent engineering, which really should be called *integrated product development* to emphasize that more than design and manufacturing engineering personnel must participate, can accelerate this portion of the development activity.

Companies must be willing to address the apparent paradox of taking more time at the front end in order to speed up the product development cycle. Previous "hurry-up, then fix" approaches must give way to an expansion of the multifunctional team effort to develop a product specification embodying the right quality to address the market's needs. Remember, it should be a *suitable* product but not necessarily the *ultimate* product. The product family approach to marketing, where subsequent, continually refined and enhanced products address the changing and evolving needs of the customer, is the only prudent product development approach to pursue.

Taking more time to get the product specification right can shorten the product development cycle.

Once the product or service is being sold and used, some level of maintenance activity will be required, as is also illustrated in Figure 1-3. This can be reduced if the product's quality is very high, but we do not know of any product or service that has been totally free of maintenance requirements. Thus, resources must be available to support this activity without detracting from development (unless your company is willing to let its reputation suffer).

Finally, Figure 1-3 also shows a post mortem or completion audit activity following development, and Figure 1-4 emphasizes this must also involve a multifunctional project team. A "failure" is not a failure unless *no* lessons are learned; a "success" is not a true success unless what worked well is incorporated into future new product development practices. In a long development project, audits may also be desirable during development at the completion of intermediate phases. The key objective of these audits is to record and institutionalize any lessons that have been learned. Determine what the company can do in the future that will simplify new product development, and make certain that procedures are changed. Regrettably, this activity is frequently skipped in the press of seemingly more urgent business. When this activity is omitted, the company is likely to repeat ineffectual behavior.

Widely disseminated results of a post mortem can avoid many product development problems in the future.

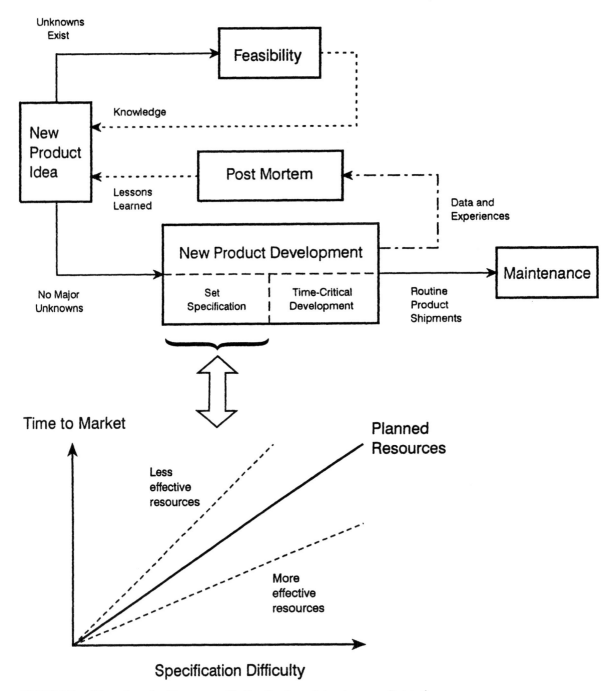

FIGURE 1-7.   Where the trade-off between specification, time to market, and resource fits into the new product development process.

TYPICAL PROBLEMS

New product development project managers and their senior management must be alert to unaddressed symptoms of external and internal difficulty. These include such problems as a consistent pattern of being later than competitors in reaching market or disharmony between various functional groups within your company. The obvious solution is to search constantly for similar shortfalls and overcome them promptly.

HIGHLIGHTS

- Successful new product development projects provide two outputs: routine shipment of the new product and lessons learned.

- In addition to the new product development itself and the post mortem, corporations must recognize that resources must also be devoted to feasibility and maintenance activities if new ideas are to be exploited effectively.

- The new product development and post mortem activities must involve multifunctional teams.

- The new product development activity requires setting the specification and following this by time-critical development, which may be divided into separate phases.

- A customer-driven quality perspective is an important element of the new product development process.

Notes and References

1. P. E. Plesk, "Defining Quality at the Marketing/Development Interface," *Quality Progress,* June 1987, pp. 28–36.
2. See, for instance, *Management Practices: U.S. Companies Improve Performance Through Quality Efforts,* Government Accounting Office, May 1991, GAO/NSIAD-91-190; *International Quality Study: The Definitive Study of the Best International Quality Management Practices, Top Line Findings,* American Quality Foundation and Ernst & Young, 1991; and M. Imai, *KAIZEN: The Key to Japan's Competitive Success.* New York: Random House, 1986.
3. M. L. Wald, "Bulb Lighted by Radio Waves May Last for Up to 14 Years," *New York Times,* 1 June 1992, p. 1ff.
4. T. C. Hayes, "Compaq Ousts Its Legendary Chief," *New York Times,* 26 October 1991, pp. Y17-18.
5. As quoted in *Focus,* February 1992, p. 2.
6. T. Cooper, "Issues & Trends in the Pharmaceutical Industry," presentation 10 June 1992 at Conference Board Chief Administrative Officers' Council, Brook Lodge.

CHAPTER 2

What Is a New Product Project?

In this chapter, we first describe how project management tools and techniques can help the new product development process and the multifunctional team. We stress that any new product development project must be considered and justified as a business undertaking, not merely a technical project. We explain seven characteristics of new product development projects and discuss their implications. The role of services is then discussed, both as the "product" itself and also as an adjunct to tangible goods. Finally, we describe the project management process and how it applies to new product development.

THE ROLE OF PROJECT MANAGEMENT IN NEW PRODUCT DEVELOPMENT

This book is about managing new product development projects—where the "product" may be either hardware or software or, increasingly often, a combination. The successful new product development project manager must be able to:

Satisfy the total product specification and quality requirements
Achieve fast time to market of a saleable product (at least for the first member of a new product family)
Live within the development budget that was initially used to justify an attractive financial return
Manage complex relationships among multiple interests, as illustrated in Figure 2-1
Handle diverse interpersonal relationships in a multifunctional project team

Commerical new product development undertakings of this sort differ greatly from many other kinds of projects, such as aerospace projects for the government (Table 2-1), capital construction projects, management information systems implementation, and so on. New product development projects where the product will serve multiple users involve complex trade-offs of myriad possible goals (that is, product specifications), schedule, and resources. Many other types of projects, often including new product development for an original equipment manufacturer (OEM), begin with a customer specifying a mandated "product," and the supplier responds with a proposed plan, schedule, and cost.

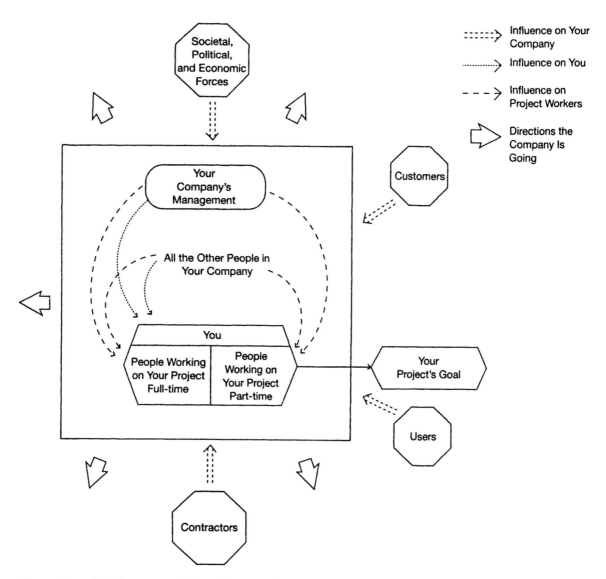

FIGURE 2-1.   Multiple company goals impede the new product project manager.

Still, a new product effort is a project because there is a goal, schedule, and budget—whether or not it is called a project. Thus, the *simple* tools of project management can be helpful in managing new product development. In what follows, we use the proven project management methodology to provide a roadmap (or skeleton framework) to describe how you can manage the development of new products and get your high-quality product or service to market quickly. Because there is a proliferation of microcomputer-based project management software, mass market and other, you must realize that

Managing a new product development project involves much more than using project management software.

TABLE 2-1.   Some Differences Between Government Aerospace and Commercial New Product
              Development Projects

	Government Aerospace Projects	Commercial New Product Development Projects
Business source	Must watch for requests for proposal, verify funding levels, and obtain contract	Market opportunity has to be identified and self-funded development is required
Time scale	Many years often required to obtain business	Short time to market is often crucial
Market research	Narrow, easily identified market	Market research is required
Market share	Typically chase several opportunities to gain one contract; that is, you win or lose rather than get some share	Typically many potential buyers, so can get some share (even if not all that is desired or in plan)
Standards	Requires significant adjunct overhead in quality assurance, documentation, contract administration, security	Company standards apply (unless market is regulated)
Procedures	Many detailed procurement procedures to comply with	Commercial standards apply (unless market is regulated)
Production and inventory	Custom production ("sell before make")	Make for inventory ("make before sell")
Quality	High quality reduces spares, repairs, and retrofit business	High quality is required and it improves profits
Profit basis	Prospective profits are a percentage of initially negotiated costs, so high costs are more profitable than low costs	Low costs are more profitable than high costs

using such software is not the same as managing a new product development project. As we point out in this chapter, new product project management involves many considerations. Because project management software is a tool that can be helpful, we provide a few illustrations of its utility in appropriate chapters; we present more extensive examples in Appendix F. It is also a trap for the unwary (as we clarify in Chapter 7).

THE NEED FOR A BUSINESS MODEL

The product financial model is an important control and analysis tool for guiding the project. The major attraction behind the new product is the financial return it will provide the company. As shown in Figure 2-2, the project financial model acts as the financial focal point for the new product. Here is where the outcome of all design, marketing, manufacturing, and support efforts gets translated into financial language for communication with the organization. Appendix D provides a complete description of how various financial measurements are calculated and used to help control and manage the development effort.

New product development projects should improve a company's profit.

The financial "answer" to a product's attractiveness (generally internal rate of return

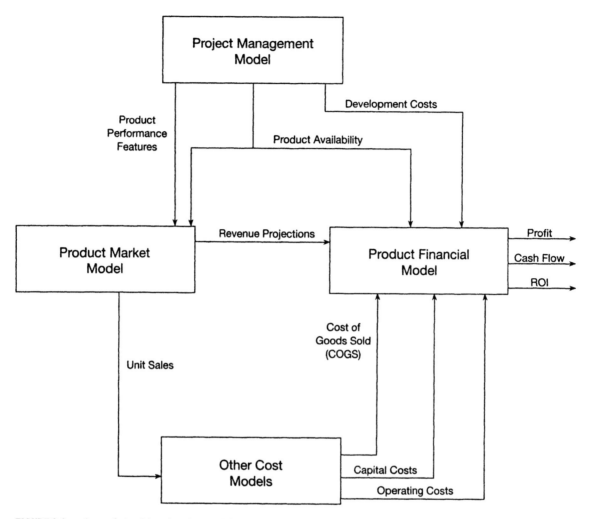

FIGURE 2-2.   Interrelationships of product models.

or net present value) is a convenient and consistent method of score keeping. Financial results play the role of a "common denominator," enabling the project manager and the multifunctional project team to add up all the changes that occur as the new product approaches the market.

DISTINGUISHING CHARACTERISTICS OF NEW PRODUCT PROJECTS

There are seven characteristics of projects that, taken together, distinguish new product project management from other managerial activities. (1) Projects may serve distinctive markets, (2) exhibit varied degrees of newness to your firm and the market, (3) have a three-dimensional objective, (4) be unique, (5) involve resources, (6) be accomplished by a multifunctional organization, and (7) vary in size. These characteristics are discussed in the following sections.

Marketplace

The marketplace can be characterized in many ways. The broadest distinction is between consumer and business-to-business (so-called industrial or commercial) markets. This applies to both tangible products and more ephemeral services. Another way to distinguish markets is as domestic, foreign, or international. Sometimes people refer to high- or low-technology markets.

The nature of the competition is also important. Once an organization or a project manager has successfully concluded a project, a high-technology product for the international market, for instance, it will be much easier to undertake a second project of that sort.

The market for which a new product project is undertaken affects how it will be done.

Newness to Company and Market

There are many ways to characterize new product development projects. Figure 2-3 illustrates one way to characterize new products, depending on newness to your company and the market. The risk is lowest for a product modification, intermediate for a line extension and me-too imitation, and very high for a new to the world product, as shown in Figure 2-4. Normally, high-risk new product development projects should be subject to more scrutiny by senior management than low-risk projects. Inherently,

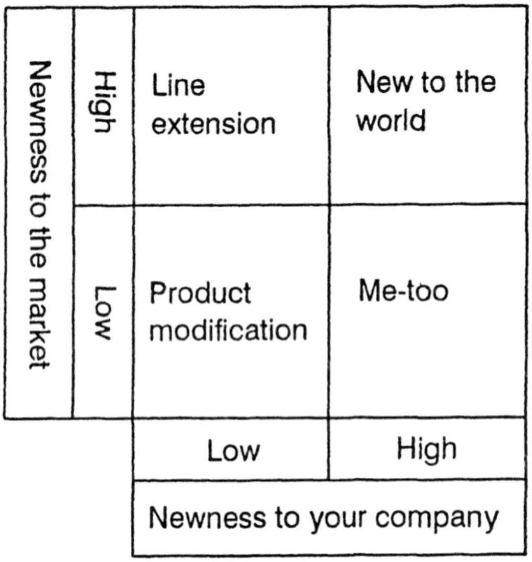

Newness to the market	High	Line extension	New to the world
	Low	Product modification	Me-too
		Low	High
		Newness to your company	

FIGURE 2-3.   Some ways to characterize the nature of new products.

a risky project will require more time to market; conversely, if time to market is critical, less product novelty and less management intrusion may be helpful. Because of these differences, the details of and phases required for a company's new product development procedure (or roadmap, as in Figure 1-3), may differ for each cell of Figure 2-3.

Figure 2-4 is not intended to be exact because the trends are merely directional indications. One company's line extension may, in fact, be quite risky. As an example, it is difficult to enter a competitive market successfully, especially when there is entrenched, high-quality competition. Two examples from the fall of 1992 provide an interesting contrast.

Compaq decided to challenge Hewlett-Packard by entering the rapidly expanding market for laser printers. Hewlett-Packard had the dominant share of this market, maintained by the introduction of a steady stream of new models and products. Hewlett-Packard had done a superb job of pursing a product family strategy with a proliferation of features and benefits.

However, a headline in the *New York Times* stated, "Hewlett-Packard may find itself playing catch-up for a change, analysts say."[1] What was going on? Compaq was attacking the standard setter by offering greater speed, more flexible connectivity, greater paper capacity—and a lower price. Compaq's only weak spots may have been print quality (but it's not yet clear that it is seriously weaker in that feature) and customer service and support (where Hewlett-Packard is outstanding).

In contrast to the rapid growth of the laser printer market, television is a very large but now slow growth market. There are about sixty million video cassette recorders (VCRs). Programming a VCR—the bane of many people's lives—has been simplified by

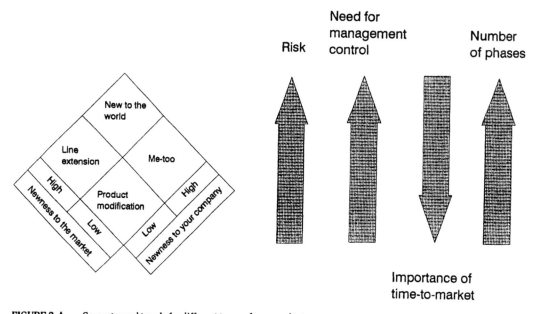

FIGURE 2-4.   Some general trends for different types of new products.

VCR Plus™, a system that provides a unique number (PlusCode™) to identify every television program. These numbers appear in the television listings of one newspaper in each major market. Viewers merely enter the PlusCode™ number of the show they wish to record in a control unit to program the VCR automatically. This capability has been available since early 1991 and impresses many VCR users as a desirable simplification. In fact, by the end of 1991, somewhat over two million VCR Plus™ units had already been sold. However, units that are sold are not the same thing as units that are used, and some purchasers may revert to conventional or on-screen programming.

However, Voice Powered Technology, a venture start-up firm (Canoga Park, California), has developed the VCR VOICE™ Programmer and Universal Remote, a hand-held controller. This is a "new to the world" product, but time to market was very important in this development. The imperative for this company has been to get its product to market before the entire market became addicted to VCR Plus™. The market is finite because it is growing slowly (compared to the expanding market for laser printers, at least), and every household that chooses VCR Plus™ is one less household that is readily accessible to VCR VOICE™. Voice Powered Technology's radio advertisement stresses the product's simplicity, and its current telemarketing follow-up sales pitch stresses "fun" and "add class to your entertainment center," all of which are hardly decisive consumer benefits sufficient to justify the price (nominally $169, but probably $149).

Voice Powered Technology also hopes to introduce products to control telephones, telephone answering machines, and microwave ovens. In its view, these products also have more features than the average consumer can effectively use.

In laser printers, it is a case of direct competition, that is, similar products in the same market. The VCR controller competition is more indirect, with product and system solutions very dissimilar to the putative consumer problem.

Thus, the type of new product is a second characteristic. A project is not an ongoing activity but rather an undertaking that ends with a specified accomplishment. The development project often ends when the company can maintain routine shipment of the new product. However, as more companies work to develop *families* of products, the overall new product development project may be viewed as a series of related projects. Many of the same team members may continue to work together on subsequent versions, but some members of the first product team may become involved in ongoing maintenance or the start of a still different effort.

A family of new products may require a related or connected series of projects involving an enlarged multifunctional project team.

Objective

New product projects have a three-dimensional objective, which is the simultaneous accomplishment of the new product's total performance and quality specification (including adjunct features), the time schedule for when routine shipments are achieved, and the development cost budget (both capital and expense). This is the Triple Constraint, and we describe it in more detail in Chapter 3. Successful project management requires that the Triple Constraint be measurable (that is, specific and verifiable) and attainable. In fact, it is crucial that the multifunctional project team that is going to carry out the new product project work know how to attain the Triple Constraint point.

A successful new product development project requires accomplishing the Triple Constraint.

Uniqueness

New product development projects originate because you (or your company) want to start selling a new product or service. Such innovative undertakings are usually the most challenging kinds of projects. Although going to work on some mornings may seem to be a major undertaking, it is not usually considered a project. Going to work is an activity that repeats a prior activity, namely, going to work the day before. Each project is unique because it is carried out only once, is temporary, and (in almost every instance) involves a different group of people.

New product development projects are one-of-a-kind undertakings.

Even though a second new product software project to write a new accounts payable system is very similar to a first such project, there will be some differences, perhaps something as simple as the format of reports. Because they differ, there is always some degree of imperfect understanding of exactly what is involved and what it will take to be successful.

Also, because projects are temporary, there are always authorization uncertainties (for instance, when a project will start and the exact scope of work to be carried out). In addition, a project does not go on forever. Thus, recruiting people to work on it is similar to staffing a company that ultimately intends to go out of business. A project starts when the first person goes to work (which is often when the product concept or idea is first enunciated) and ends when the last person's work is finished (which may be at the time of product launch or sometime shortly after that when, for instance, a post mortem is completed). Somewhere in between the start and finish, several or many people may be involved. In the case of *families* of new products (for instance, versions of Boeing's 747 aircraft or Hewlett-Packard's laser printers), the project may last for more than a decade.

New product development projects involving families of products may last many years, whereas others may end quite quickly.

Finally, the people who work on one project are rarely those who have worked together on a previous project. A multifunctional project team is normally composed of people largely selected by different department managers rather than the new product development project manager. Thus, during the new product development effort, friendships or antipathies are being created with significant consequences. Strong friendships may make people reluctant to end a project because they will no longer be working together. Conversely, antipathies may make it extremely difficult for two people to cooperate during the project.

Resources

Projects are accomplished by resources, namely, people and things. Many of the required resources are only marginally under the effective control of the new product project manager. For example, a required lathe may be controlled by a model shop group, or a required computer may be controlled by a data-processing group.

The project manager must organize the correct human resources to take advantage of the available physical resources. Then the project manager has to deal with the constraints and emotional problems inherent in their use while trying to accomplish the technical performance goals within the development schedule and budget. Manag-

Managing new product development projects means managing people.

ing people is often the most difficult aspect of managing a project, especially for recently appointed managers.

If new product project managers have a technical background, for instance, engineering (which is not uncommon), there is a special challenge. They have to avoid the technical expert's propensity to concentrate on the quantitative aspects (such as engineering analyses or task budgets)—although these are not unimportant—and instead become more oriented to making things happen through people. It is a difficult challenge for a technically trained person to manage the efforts of engineering (or other technical) people. However, new product development requires the project manager to make things happen through people in other functions, marketing and manufacturing, for instance, where the personal orientations and loyalties may be vastly different. Many good technical experts make poor project managers because they can't deal with the intangibility of people issues, such as the need to "sell" (and resell) a project to other managers. Some project managers recognize this need but cannot communicate effectively.

Resource conflicts are inherent, as illustrated in Figure 2-5, which shows only two projects. Their need for a particular resource (square feet, hours of usage of a DNA

PLAN

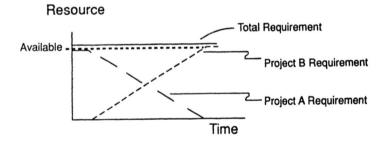

ACTUAL

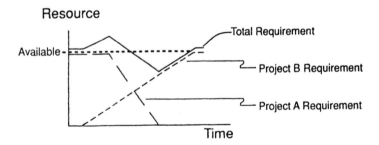

FIGURE 2-5.   Resource conflicts are inherent because the actual project resource requirements never produce a work load (on people in general, people of a particular skill, or facility) that perfectly matches resources available.

synthesizer, personnel in general, or a specific kind of human resource, for example, junior level analytic chemists) is shown. A planned requirement exactly matches the available resource. However, the timing of one project is slightly altered, and as a result, the required resource no longer matches what is available. First there is too much work, and later there is not enough work. Resource overloads and underloads are common—and bedeviling—characteristics of the new product development project manager's world.

These mismatches are bad enough where the resources are physical things. The resulting problems are often much worse where the resource mismatch (be it overload or underload) is people.

Because of the tremendous importance of these people issues, we thoroughly discuss organizational options and their influence on resource control and availability in Chapter 10. The rest of Part 4 is devoted to an extensive review of effective actions you can take as project manager to lead both the project and support teams.

New product development project managers must spend their time with people.

Organization

Every organization has a multiplicity of purposes at any given moment, if for no other reason than it is composed of many individuals with varied skills, interests, personalities, and unpredictabilities. Thus, the project manager will often be frustrated by the many other directions in which the organization seems to be (and often is) moving. These multiple directions arise because of personal aspirations and interests, because of various parochial interests by different components of the organization, and because of many projects being carried out simultaneously.

Every new product development project is subject to many influences.

As Figure 2-1 suggests, customers and users (who may be your customers, customers of your customers, or other people) will also affect your organization. Thus, in many instances, the project manager (even with a loyal and dedicated team) may be frustrated in trying to achieve the project goal.

New product project management, in large part, is the management of interpersonal conflict, which is inherent in complex organizational situations.

Size

"Program" is commonly used synonymously with "project." Thus, the expression "program management" is often used interchangeably with "project management." Some organizations use "task management" as well. Program management, project management, and task management are generally identical. But programs are usually larger than projects, and projects are usually larger than tasks. Thus, there is some connotation of size when terms other than project management are used. Nevertheless, the techniques and methodology are essentially the same, differing only in detail. We use "project" throughout the rest of this book.

A new product development project may be of any size.

SERVICES

Services are different than "goods" or tangible products. We are not just describing the services that go along with the new product, such as support, maintenance, or training.

Services are a major product area in their own right. Much has been written concerning the evolving U.S. "service economy," and the value of services in the gross national product now exceeds the value of manufactured goods.[2] Indeed, services are a current battlefield of international competition. National boundaries used to provide some degree of protection for services, but that time has passed, as can easily be seen in the U.S. banking industry.

Although most of the principles outlined in this book apply equally to goods and services, there are important differences. The "goods" product developers can learn

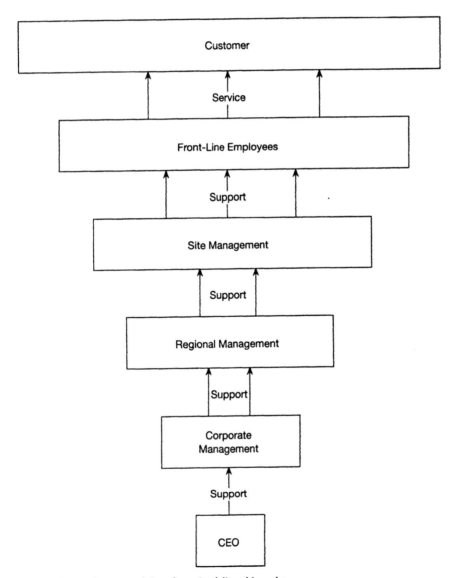

FIGURE 2-6.   A conceptual view of a service delivery hierarchy.

some important lessons from the services end of the business. If a manufactured product provides a *service* to the customer, then a service provides both a benefit and a feeling. Because the service provider is frequently a person, the technical capability and attitude of that individual to a large degree become the product.

But we cannot overlook the *process* behind the service delivery. For example, the company must emphasize the importance of making investments in technology and training to *support* rather than monitor the customer contact process.

The process for developing new services is different than that for goods. Generally, the service process is much faster and simpler. Unfortunately, the lower barriers to entry and fewer restraints belie an even greater need to keep a focus on customer needs. Indeed, it is suggested that to guide the evaluation of a new service, three strategy elements need to be emphasized throughout the development process:

Market research—do we continually get feedback from the customer for the service?
Business mission—does the new service fit with our basic business and our target
    markets?
Organizational values—how do the guiding principles and values of the business fit
    with the new service?

The well-organized service organization has been defined as an inverted pyramid, as shown in Figure 2-6. Customers are on the top, and a customer's contact with the front-line employee is the critical interaction, the place where value and quality are determined. The rest of the organization, right down to the chief executive officer (CEO), is there to provide the support and tools to make that interaction the best possible.

WHY NEW PRODUCT DEVELOPMENT IS DIFFERENT

When considered in the range of project management activities, the new product development effort has many unique characteristics. Unlike a project to build a bridge, design a microprocessor, or write a book, new product development is a *business* rather than a functional undertaking. When the new product development project is concerned with an entire family of products, it may become a freestanding business unit within a larger corporate entity. The multifunctional project team in this situation may be the entire divisional staff. As such, the new product development project is really a "super" project management task coordinating the timing, budget, and output of interlocking functional projects. Also, the market and competitive uncertainties mean that success is not totally under project control.

The uniqueness of the new product development project can be summed up in three areas: the product specification, the customer, and the multifunctional team.

The Product Specification

One of the more difficult balances in business is developing a product that is new and innovative yet utilizes the existing strengths and capabilities of the firm. Overall, the new product specification, its timing, distribution, and support, must fit with the direction in which the company is headed—the corporate strategy.

Corporate strategies usually define products that complement existing capabilities in production, distribution, marketing, and support. The new product should also fit the "position" the firm occupies in the market. It does a firm little good to develop a technically superior product that can't be sold through existing channels, is incompatible with existing products, and can't be supported by the existing service organization. It is also difficult for a firm that has positioned itself as the "Cadillac" in its market to offer a new product with low price as a major attribute. In the late 1980s and into the early 1990s, for example, both IBM and Compaq (with "Cadillac" products) were unable to compete effectively with manufacturers of low-price clone computers.

The new product specification represents a careful balance between technical risk, market requirements, and corporate capabilities.

The Customer

The major influence for the product specification, support, pricing, and so on is the customer. Customers get a very important vote when it comes to the product; that vote is called a product *purchase*. The mixture of benefits your new product is bringing to the marketplace must meet the customer's needs better than your competitors. If it doesn't, you simply don't get many "votes." In the end analysis, the customer is the ultimate determination of the success or failure of the new product development effort. Sometimes it is easy to lose this very important perspective.

A crisp definition of the "customer" is a vital but often diffcult concept to pin down. Sometimes considerations of market segmentation, distribution channels, and product interdependence can complicate the effort. We address these considerations in later chapters.

The customer is the ultimate determination of a new product's value.

The Multifunctional Project Team

As mentioned previously, the new product development effort is actually a superproject management task—coordinating and managing the subprojects of functional team members. The importance of the multifunctional team in delivering customer value cannot be overemphasized.

Many times, customer support, pricing, application development, or distribution effectiveness can more than compensate for shortfalls in product performance. Remember, your customer is really buying a bundle of attributes (from a large multifunctional organization, your company), and the importance of different product elements can vary widely. This is why blending the perspectives of the multifunctional team is so important.

THE "GLUE" OF THE NEW PRODUCT DEVELOPMENT PROJECT

The glue that holds the new product development project together is the business plan. The plan is a document that tells how the development functions will work together to design, produce, sell, and support the new product. It is the new product development project manager's bible or script, documenting the tasks, assumptions, timing, risks, and payoff for the effort.

As we said before, the core of the new product development business plan is a financial model of the new product's projected performance. The financial model is the "quantification" of the assumptions made in the business plan. The payoff is indicated in the project's projected financial results. Basically, it is the "answer" or the reason for the undertaking.

In its simplest form, product profitability is the parameter used to guide many critical new product development priority and trade-off decisions. The financial return on investment serves as a common denominator that allows for an objective evaluation of the many decisions that must be addressed during the new product development project.

The business plan and financial model tell the new product story and serve as an important reference document for the team.

THE NEW PRODUCT PROJECT MANAGEMENT PROCESS

Project management requires five different managerial activities and can thus be most simply structured as a five-step process:

1. Defining: defining the project's goals
2. Planning: planning how you and your team will satisfy the Triple Constraint (goal) of performance specification, time schedule, and money budget (The plan depends on the mix of human and physical resources to be used.)
3. Leading: providing managerial guidance to human resources, subordinates, and others (including subcontractors) that will result in their doing effective, timely work
4. Monitoring: measuring the project work to find out how progress differs from plan in time to initiate corrective action (This often leads to replanning, which may force a goal [definition] change, with a consequent need to change resources.)
5. Completing: making sure the job that is finally done conforms to the current definition of what was to be done and wrapping up all the loose ends, such as documentation

Figure 2-7 illustrates where these activities logically fit into the new product development process roadmap. The first two steps are not necessarily separate and sequential, except when the project initiator issues a firm, complete, and unambiguous product specification that the development team feels confident it can accomplish. This might be the case where the new product is being developed primarily for an OEM.

The five managerial activities fit logically into the roadmap.

It is more common to start the development work with a proposed work definition (that is, a tentative specification), which is then jointly refined after preliminary planning elucidates some consequences of the initially proposed work definition. Specifically, the multifunctional project team must make trade-offs between the specification's difficulty and time to market while giving realistic consideration to the resources that will be available, as illustrated in Figure 2-8. In distinction to new product development for an OEM, the specification will have to satisfy many buyers. The final definition must be measurable (specific, tangible, and verifiable) and attainable (in the opinion of the people who will do the work) if you want to be successful.

Thus, in fact, the resources to be dealt with in the leading phase must be considered before planning can be finished (see Figure 2-8 [A]). For instance, you might need

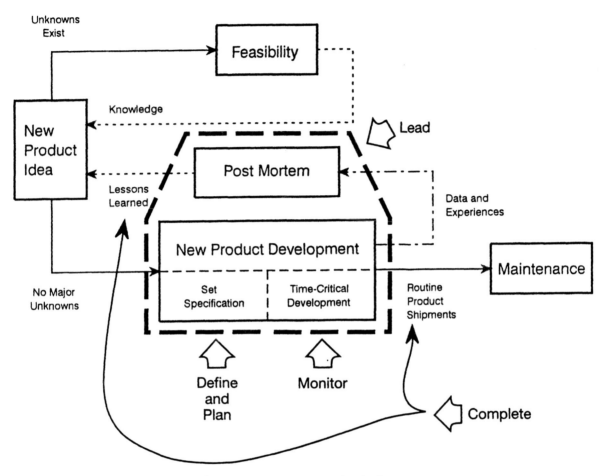

FIGURE 2-7.   Role of the five project management activities in the new product development process roadmap.

engineers familiar with carbon fibers if the new product is to be a high-performance golf club, whereas you would use a metallurgist if the product was a conventional golf club.

No project goes in accordance with your plan. What you don't know when you start is where it will go awry. Consequently, as later chapters illustrate, replanning is almost always required, sometimes leading to a change of the definition (see Figure 2-8 [B])—in this case, the product specification (which, unfortunately, almost always delays the development project's schedule and risks missing the market's window of opportunity). Ultimately, the project can be completed when the work that is done satisfies the current specification requirement (see Figure 2-8 [C]), and routine product shipments commence. (If the new product is merely the first member of a planned family, the same multifunctional project team may make a transition to work on subsequent family members. However, the development of each member should be a separate project, even if linked. In such a

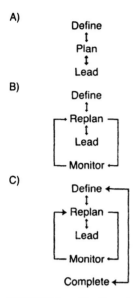

A)

Define
↕
Plan
↕
Lead

B)

Define
↕
→ Replan
↕
Lead

Monitor

C)

Define ←
↕
→ Replan
↕
Lead

Monitor ←

Complete ←

FIGURE 2-8.   The five activities are interdependent.

case, some projects or family members may not be financially attractive on their own, but the obvious goal is to be certain that the total family is an attractive business.)

Nevertheless, the five-step managerial activity process covers each required action and is a useful conceptual sequence in which to consider all project management. Thus, we organize our discussion of the management of new product development projects according to it.

Projects require five managerial activities, which may overlap.

The Triple Constraint, an extremely important notion for all project management, provides the defining parameters of any project. In the specific case of new product development projects, it consists of items such as the following:

I. Product attributes (those things seen by a customer or user)
   a. Performance features
      1. Speed
      2. Information display
      3. Capacity
      4. Sensitivity
      5. Resolution
      6. Weight
      7. Size
   b. Adjunct features
      1. Warranty
      2. Product documentation
      3. Telephone support
      4. Service

      5. Maintenance arrangements
      6. Payment terms
      7. Delivery
   c. Quality
      1. Ease of use
      2. Durability
   d. Total Cost
      1. Acquisition cost
         (a) Sell price
         (b) Sales taxes
         (c) Shipping
         (d) Installation and setup
         (e) Training
      2. Operating cost
      3. Maintenance cost
  II. Development schedule (from idea to market entry)
 III. Development budget
   a. Expense
   b. Capital

All projects are defined and characterized by a Triple Constraint. In addition, new product development projects are the initial part of the inception of the product life cycle. The PLC entails four stages (innovation, growth, maturity, and decline) and provides an additional framework for evaluating and justifying new product development projects.

Note that money appears in two dimensions of the Triple Constraint. The *product attribute* of total cost (which determines your selling price and gross margin) is a major factor for the prospective buyer, and its acceptability depends on the other product attributes. A substantial user benefit can justify a higher price. The development budget (and the actual costs incurred during development) is important in the *project management* process. The development budget has to be low enough so that the product development effort has an attractive payback to your company. We describe various discounted cash flow measures to assess financial payback and return on investment in Appendix D.

Money appears in two dimensions of the Triple Constraint.

TYPICAL PROBLEMS

Failure to identify a new product project for what it truly is usually leads to missed specifications, late completion, and/or a budget overrun. The solution is to recognize that there is a project when something must be done and then to organize to complete the project in the least disruptive way.

The fact that humans are involved in projects and must be worked with is often especially troubling for technically trained project managers. In many cases, this need to work with people is the biggest obstacle that a good technologist encounters. Such technically trained managers expect, but do not get, completely logical or rational behavior from these people (or, for that matter, from themselves). Finally, although many project management tools (for instance, many of the planning and monitoring tools discussed in this book) are completely rational, project management in an overall sense is not an exact science.

HIGHLIGHTS

- New product projects are temporary undertakings with a specific objective that are accomplished by organized application of appropriate resources, especially a multifunctional project team.

- Seven significant characteristics of new product projects are the newness to the company and market, the nature of the marketplace, the three-dimensional objective, uniqueness, the involvement of resources, interaction with the organization, and size.

- Size and complexity do not distinguish projects from other activities.

- The Triple Constraint defines a project's goal: performance specification, time schedule, and money budget. This goal must be consistent with the company's total business and profit plans and must satisfy market requirements.

- New product project management is the process of achieving the objectives in any organizational framework despite countervailing pressures. This often requires selling and reselling others on the project's importance.

- There are five managerial activities in new product project management: defining, planning, leading, monitoring, and completing.

- Successful management of a new product development project means shipping a product that meets the performance specification, on schedule, within the development budget so as to satisfy the corporate business and profit plans.

Notes and References

1. P. H. Lewis, "Compaq's Bold Plunge into the Laser Printer Market," *New York Times*, 13 September 1992, p. F9.
2. See, for instance, L. A. Schlesinger and J. L. Heskett, "The Service Driven Company" *Harvard Business Review*, September-October 1991, pp. 71-81; C. W. Miller, "Growth Forum: Crossing the Line—Going from Goods to Services in the Product Development Process," *Visions*, September 1992, pp. 10-15; and K. Albrect and R. Zemke, *Service America! Doing Business in the New Economy*, Homewood, IL: Down Jones-Irwin, 1985.

Defining the Goals of a New Product Project

CHAPTER 3

The Triple Constraint

This chapter introduces the concept of the Triple Constraint as a project definition, identifies some of the obstacles to satisfying it, and describes some steps to help achieve it. The consequences of various project outcomes are considered from the point of view of satisfying the Triple Constraint. Next, we point out that the new product development project manager has to satisfy both external and internal customers. Finally, we discuss the tricky balance between encouraging creativity while maintaining management control and introduce the use of a staged (or phased) process as part of the new product development activity.

THE CONCEPT OF THE TRIPLE CONSTRAINT

Figure 3-1 illustrates the Triple Constraint, a very important concept we emphasize throughout the book. Successful project management means achieving all the product attributes (such as those illustrated in the list on pages 33–34), which derive from setting the specification, as illustrated in Figure 3-2, on or before the time limit and within the budgeted cost. The cost is usually measured in dollars, francs, marks, or whatever the coin of the realm but may sometimes be measured in the number of labor hours, or perhaps labor hours in each of several labor rate categories such as senior research chemist and laboratory technician; system analyst and programmer; senior engineer, engineer, and junior engineer; and so forth. The key point the Triple Constraint illustrates is the need to satisfy simultaneously three independent goals—not just one.

The Triple Constraint defines projects: performance specifications, time schedule, and money (or labor hour) budget.

OBSTACLES TO SATISFYING THE TRIPLE CONSTRAINT

Unfortunately, the Triple Constraint is very difficult to satisfy because most of what occurs during any project conspires to pull the performance below specification and to delay the project so it falls behind schedule, which usually makes it exceed budget. Because no project follows the plan, the successful project manager must be alert to potential problems to achieve the Triple Constraint.

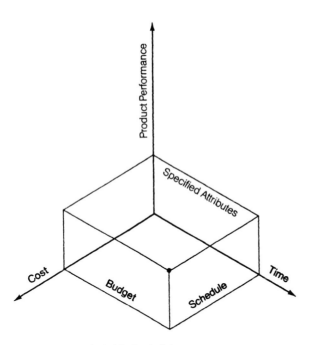

FIGURE 3-1. The Triple Constraint.

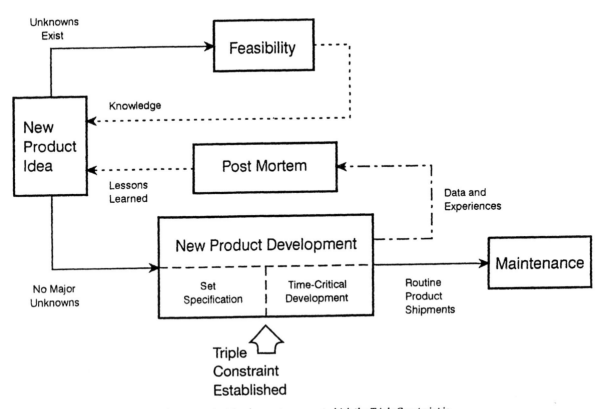

FIGURE 3-2. The point during the new product development process at which the Triple Constraint is established.

Under the best of circumstances, it is difficult to achieve the Triple Constraint. However, it is also normal for the Triple Constraint to change during the course of the project. For example you or the project team may become convinced the original Triple Constraint cannot be achieved and will have to propose an alternative end point. A change may be imposed by, for instance, new legislation or regulation. A more pervasive matter is that the varied members of the multifunctional project team will almost surely have differing perceptions on many issues.

As you can see, projects encounter a wide variety of problems. Some of the principal ones are enumerated in the following sections, organized by the dimension of the Triple Constraint most affected. Sometimes only general symptoms are evident, rather than specific problems with a single dimension of the Triple Constraint. Although Figure 3-1 shows these three dimensions to be orthogonal axes, hence mutually exclusive, project management is not that ideal.

Product Performance Problems

There are many reasons the performance specification is difficult to achieve. We discuss three of them. First, there may have been poor communication between members of the multifunctional development team. That is, they have different perceptions of the specification, or the wording is ambiguous. For instance, "security" means different things to different people or in a different context. To a military contracting officer, it may mean a secret classification, whereas a person working on a military software project might interpret it as meaning data protection. At the Social Security Administration, the word may mean enough money to live well, whereas in the brokerage industry, it might mean stocks or bonds. In short, capable, dedicated, and perfectly honest people may interpret a simple word differently—with attendant problems. Product specifications must not be simply an exercise to cram maximum performance into a limited space or per dollar of selling price. Integrity is an important concept that views the product from the customer's standpoint.[1] For example, in 1987, Mazda put a high-performance, four-wheel steering system into its family five-door hatchback. Honda put the same system into its sporty Prelude. Not surprisingly, the Mazda system sold poorly.

A winning product is not one that simply "outfeatures" the competition. It often has an important element of consistency or integrity from the customer's viewpoint. Two important ingredients foster this consistency:

Early and frequent customer input throughout the new product development cycle
A "heavyweight" (that is, senior and experienced) product manager who has the
    maturity, vision, and power to orchestrate across functions

Today, quality is generally viewed as satisfying users' needs by correctly defining their requirements. We consider quality—however defined—as being a part of the product attributes. Poor or ambiguous communications may lead to a quality disappointment. This quality shortfall can occur even if other performance specifications are well satisfied. For example, a "good quality finish" on a product's surface may not be precisely measurable, and the result may be either a disappointment or unacceptable.

A second problem arises because assumptions have been too optimistic. Goals may have been too ambitious, which is not uncommon in advanced technology.

Third, the development team may do a poor design job or make mistakes. Unfortunately, workers (and managers) make errors occasionally, and these errors may cause a performance deficiency.

Clarify unclear specifications.

Development Time Problems

Schedule problems arise for several reasons, the most insidious being an overemphasis on the product's performance at the expense of a balanced view of the Triple Constraint. For instance, scientists or engineers, who are commonly appointed project managers, tend to concentrate on the technology and to strive for technical innovations or breakthroughs. A computer programmer may emphasize work on a clever algorithm or use a new programming language rather than expeditiously completing the program with existing capabilities. Such striving is accomplished at the expense of the schedule, and it frequently has unfavorable cost repercussions. To put this another way, "better" is the enemy of "good enough."

Even where a fascination with technology is not overwhelming, technically trained people, such as engineers, tend to assume the technical performance specification is sacrosanct, whereas they consider it permissible to miss the schedule or budget objectives. Conversely, a customer might be satisfied (if not ecstatic) to be able to buy your product with 90 percent of the development team's intended product attributes, provided it is available in a timely way. And your company management will also appreciate this, especially if the team has not exceeded the development management budget. One of the many virtues of trying to develop entire product families is that enhanced features and new technology can be embodied in later members of the family. This permits the company to get a simple version to market quickly to establish a leadership position (assuming, of course, that the simple model has real—if limited—functional utility at a fair price). Members of the multifunctional development team who want to change the product specification can often be dissuaded from the "better is the enemy of good enough" tack when they are confident they can put their pet ideas into later models.

Technical excellence usually interfaces with meeting the schedule.

Figure 3-3 illustrates various Triple Constraint outcomes and their relationships. For example, performance that is better than the specification is normally obtained only if there is a budget overrun or a late delivery, and most frequently when both occur. This illustrates why it is so important for the project manager to moderate the enthusiasm of technical experts, who can always see the possibility of doing better and are eager to achieve higher performance. Note that although we talk about *a* product performance goal, this is itself multidimensional, as was illustrated in the outline list at the end of Chapter 2. That is, there are typically many product features, adjunct features, quality aspects, and the purchaser's total cost of acquisition to consider. Thus, in the real world, the new product development effort may produce better product performance than the specification in some regards and worse in others. This reality obviously complicates the simple model shown in Figure 3-3. What ultimately matters is how potential buyers view the achieved product performance that is actually achieved.

A second source of difficulty in meeting the schedule arises because resources are not available when required. These resources may be either equipment (such as lathes

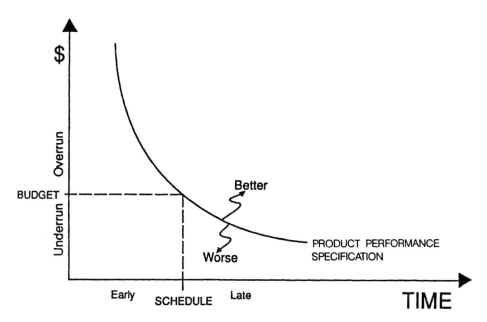

FIGURE 3-3.   Triple Constraint outcomes.

or computer hours) or personnel (a well-qualified circuit designer, for example). This absence of planned resources forces the project manager to find substitutes, which may require a subcontract to get some design work done. Or it may mean using marginally qualified people who take longer to complete a circuit design than the well-qualified person previously assumed to be available.

Third, a project can get into schedule difficulty because those assigned to it are not interested in their tasks. In this case, they may choose to work on other things or work halfheartedly on the project.

Fourth, schedule problems can arise because the performance specification is altered. For instance, increased efforts that lead to additional work are accepted. A common occurrence that illustrates this is a salesperson asking for a few extra control switches. The project manager may misperceive this as trivial because a control panel is already being provided. If he or she agrees to provide these extra switches, which were not part of the original specification, the project manager is agreeing to do additional work (perhaps even modifying the control logic) without changing the schedule (or cost). There is, however, additional work called for to put in the switches, and it does not take many changes of this sort first to produce a one-day schedule slippage and then a one-week and so on until the project is in serious schedule difficulty.

The time schedule should change if the performance specifications change.

Development Cost Problems

Cost problems arise for many reasons. When a project is in trouble on its time dimension, it will often be in trouble on its cost dimension as well because resources are not being used as efficiently as planned.

A second cause is the "liars contest" that often occurs inside an organization. In this case, the project has to be sold to upper management, and you are competing with other organization managers for authorization. Sometimes this competitive pressure induces the development team to be foolishly confident.

A third source of cost difficulties arises because many of the initial cost estimates are simply too optimistic. They do not reflect the inefficiencies that will occur in scheduling resources to perform the work or the fact that less well qualified people may be assigned to do the work.

Occasionally, mistakes are made in the cost estimating. Like design mistakes, these are unfortunate, and careful scrutiny and review can minimize this occurrence.

A fifth reason for cost problems is simply an inadequate cost consciousness on the part of the project management or a failure to have an adequate cost management system. This is never excusable.

THE NEW PRODUCT DEVELOPMENT
PRODUCT MANAGER'S TWO CUSTOMERS

New product development project managers have two distinct "customers": the obvious external customer in the marketplace and an important customer in the sponsoring organization. They must keep both in mind as they navigate the way toward product launch. The characteristics and demands of these two customers are frequently very different, quite often in direct conflict.

There are both internal and external customers.

The External Customer

Although the external customer for the product is more obvious, there are frequently several aspects that must be considered to get a clearer understanding of "the customer."

Market Segments

Different segments can differ considerably in their requirements and uses of the product. It is important to understand that the market for products can be segmented along many dimensions. Typical segments could be size of the firm, industry, functional department type, or existing customers. In other cases, it may be useful to group customers according to their behavior or potential application for the product, such as innovators, early adopters, price sensitive, or heavy production users.

The product specification must be in sufficient detail to define technology, performance, and manufacturing requirements. However, to be successful, it must also contain a healthy dose of imagination.[2] Companies need to include imagination to lead rather than just respond to customer needs. A matrix of needs, applications, and functions should precede the matrix of customers and products.

Companies also need to explore markets with alternative products. Betting on a "home run" every time at bat is not only unrealistic, it is counterproductive. Toshiba, for example, introduced over thirty different laptop computer models between 1986 and 1990. This avalanche of products guaranteed there would be failures, but it also guaranteed Toshiba's successful dominance of the rapidly expanding market for laptop computers.

The ability to be creative and innovate in anticipation of market needs often requires a redefinition of "failure." We need to redefine the line between a "product" and the beta test (initial field trial) of a concept. Sometimes stretching the limits and getting the feedback a "failure" provides is the only way to a successful new product development.

Being creative and innovative may require redefining "failure."

Distribution Channels

Many firms frequently define as customers organizations that either distribute or integrate the product as part of a larger product or system of products. Distributors (including, for instance, retailers and manufacturer's representatives) are important "customers" of the firm. It is important to recognize, however, that they are not the end user of the product. It is vital to understand your customer's customer. The end user or consumer is the ultimate determination of a product's value.

Market segmentation and product distribution requirements can frequently complicate a simple understanding of the new product's customer.

The Internal Customer

To a very real extent, sponsoring corporate management is a customer of the new product development effort. The new product development project is frequently in competition for the scarce resources of the firm. Most organizations cannot afford the funds and labor to support all the new product development projects presented. They must make project, timing, staffing, and funding decisions that are in the best long-term interests of the firm. Frequently, new product development program reviews are an opportunity for new product development project managers to resell their programs to management.

Another important customer of new product development project managers is the multifunctional team member. Project managers must win the dedication and motivation of the team members. They must employ recognition and incentives to win team members from their natural inclination to give first allegiance to their functional areas.

The new product development project manager must be sensitive to the importance of internal "customers" for the new product.

A Perspective

What we are describing are project managers who are obvious product advocates. But project managers are really in somewhat of a paradox. If they are simply product advocates, effectiveness will be diminished. They are expected to keep the firm's broad objectives in mind. They must make the case that the product decisions make good business sense for the firm and have the vision and perspective of a good businessperson in bringing both enthusiasm and objectivity to the task.

THE PRODUCT DEVELOPMENT PROCESS

The new product development process presents many firms with a paradox. New product development is an area where the skills of creativity and the entrepreneur are valued. Yet the expense and importance of new product development to the firm's future dictate close monitoring and control. Staged product development is the essence of a formal new product development process and is an attempt to address this paradox— creativity with discipline. The experiences of many firms indicate that a formal stage gate approval process provides both the structure to evaluate new ideas and the mechanism to speed those ideas to the marketplace.

However, any new product development procedure, staged or not, is merely a framework. No procedure will assure that a new product truly meets market needs. It can reduce the likelihood that a company will not overlook steps and actions to promote that desired result. As such, a good procedure provides a focus for development work, clarifies what must be done, when, and by whom.

The objective of the new product development process is to encourage creativity, resulting in the rapid development of profitable new products fo the firm. Companies typically have an excess rather than a scarcity of new product concepts from which to choose. They need a fast, simple, yet effective way of screening out less attractive alternatives so the concepts presenting a higher probability of success get adequate resources. This screening process is simple yet difficult. To evaluate the fit with existing products and markets is not enough. The screening mechanism must accept change—a vision of where the market and technology are going. Things are simply moving too fast, upsetting our existing paradigms, to extend existing trends. We must be willing, even anxious, to push new products beyond the existing market-technology envelope in which we currently operate. Firms must address the difficult balance of profit, innovation, and change. We must not prepare new products for today's customers and markets, but anticipate the marketplace as it will look tomorrow.

Staged product development is an effective process to screen concepts and provide the necessary resources and structure to bring quality products to market faster.

New Product Development Stages

Firms with a formal new product development process generally turn ideas into marketable new products more efficiently and faster than those without such a process. The new product development staged process is typically customized to meet the individual needs and culture of the individual firm. The following examples contrast the staged product development approaches used by six firms.

Motorola[3]

1. Product definition
2. Contract development
3. Development through manufacturing start-up (team defines substages)
4. Program wrap-up (learning)

Kodak[3]

1. Customer mission/vision
2. Technical demonstration
3. Technical/operational feasibility
4. Capability demonstration
5. Product/process design
6. Acceptance and production

General Electric[3]

1. Customer needs
2. Concept

3. Feasibility
4. Preliminary design
5. Final design
6. Critical producibility
7. Market/field test
8. Manufacturing feasibility
9. Market readiness
10. Market introduction follow-up

Xerox[4]

1. Preconcept
2. Concept
3. Design
4. Demonstration
5. Production
6. Launch
7. Maintenance

Calcomp[5]

Specification
   1. Market requirement
   2. Specification
Development
   3. Design
   4. Engineering model
   5. Prototype
Verification
   6. System verification
   7. Manufacturing verification
Manufacturing
   8. Production
   9. End of life obsolete

Exxon Chemical's polymers innovation process[6]

1. Idea
2. Preliminary assessment
3. Detailed assessment
4. Development
5. Validation
6. Commercial launch
7. Postlaunch review

One observer stated that phased approaches generally are constructed like this:

Get an idea
Prove it works

Develop and test it
Scale it up
Launch[7]

Keithley Instruments uses seven phases, as illustrated in Figure 3-4.[8] The first three require management approvals prior to starting the subsequent phase. Also, each phase has detailed end-of-phase deliverables and requires that specific "gateway documents" be provided before the phase is judged complete (two of which are indicated in the figure). To assure that this occurs, there is a matrix indicating actions for which each functional department is primarily responsible.

In our terminology, most phases of staged development will occur during the time-critical development portion of the new product development activity, as shown in Figure 3-5. As already indicated, in many staged processes, specific criteria are clearly designated as prerequisite for the end of stages. The following list summarizes some typical end-of-stage deliverables that may be required to proceed to the subsequent phase:

Product phase	Deliverables
Concept phase	Define customer requirements
	Complete economic and technical feasibility analysis

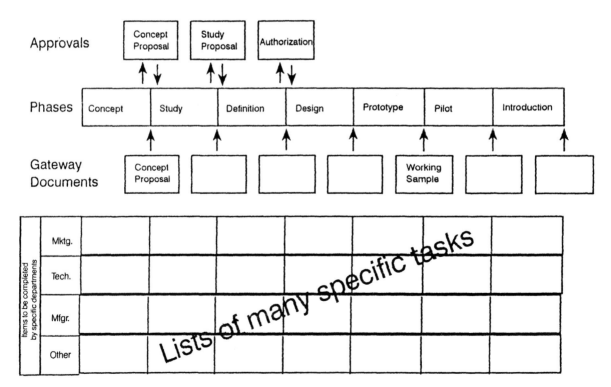

FIGURE 3-4.   Overview of selected elements of the new product introduction process used by Keithley Instruments.

	Identify critical success factors
	Develop phased product plan
Development phase	Confirm business viability
	Select technology
	Develop functional team plans
	Initial financial projections
Design phase	Benchmark product concept
	Test product functionality
	Demonstrate manufacturability
	Beta test prototype product
Manufacture/launch phase	Verify customer acceptance
	Confirm field readiness
	Ramp up production process
	Transfer to current product

This practice is quite common and very helpful in focusing the attention of the multifunctional project team. Note the overall similarity in all the illustrated processes.

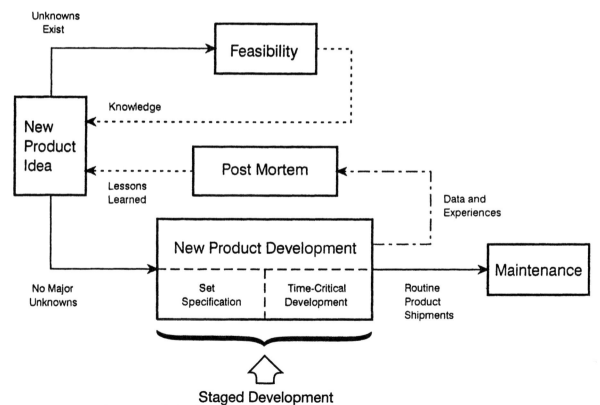

FIGURE 3-5.   Where stage (or phased) development normally fits into the new product development process.

Although individually quite different (as we have illustrated), most approaches contain the following elements: a concept phase, a development phase, a design phase, and the manufacture and launch phase. We discuss these in the following sections.

Concept Phase

This initial product phase is concerned with changing an idea into a product concept. At this stage, it is important to define the boundaries of the concept rather than the details. What market will the product address, who are the customers, their applications? What are the major technical requirements? How good a fit is the concept with manufacturing capabilities, distribution channels, and existing products?

Often these issues and others are addressed in a "concept"-level business plan that documents critical assumptions, risks, opportunities, and milestones required to exit the concept stage of development.

At this early stage, it is important to include the customer as a member of the product development process. Customer feedback can be critical in providing insights into how potential customers will use and evaluate the new product.

Customer feedback is vital in the concept phase.

Development Phase

In the development phase, attention turns toward the product specification, initial plans by functional team members, and major technical and marketing unknowns. During this phase, it is important to set firm product goals and make the basic decisions and trade-offs of product features, options, and other capabilities. Customer-driven QFD efforts are an important facilitator of this effort.

Functional team members cooperate to address technical risk factors and anticipate product manufacturing and launch plans. Preliminary models of the concept are developed to provide a "live" model for preliminary market testing and manufacturing and service assessments. A firm product budget and schedule are put in place. The development phase involves a major agreement between marketing and engineering that the product as specified will meet the needs of the market.

Firm budgets and schedules are created during the development phase.

Design Phase

The design phase is where the concept takes a major step toward reality. Product specifications have been brought to a point where preliminary "preproduction" models of the product are available. Manufacturing soft tooling is in place along with a demonstration of the capability to build to product specification.

Initial "beta test" machines have been placed with customers in order to obtain feedback on a wide range of capabilities, including:

- Product performance
- Product reliability
- Customer applications
- Product documentation and customer training
- Service and support

Manufacture/Launch Phase

This final phase is when the new product becomes an existing product. The major impacts are to the manufacturing, sales, and support organizations. Product specifications have been released to manufacturing. The field sales force is trained. Service personnel are in place and trained, and spare parts are available. The product starts to fill the pipeline.

The new product comes into existence during the manufacture/launch phase.

It is important to include elements of continuous improvement as the product exits the new product development process. These include the following:

- A project post mortem should evaluate the development process. Areas of exceptional performance as well as candidates for improvement should be documented.
- Feedback from actual customers is obtained to track the overall level of satisfaction, applications, and product performance shortfalls and surprises.

These follow-up and feedback activities contain the necessary elements for both continuous improvement of the process and the development of follow-on products.

All members of the multifunctional team should actively participate in these new product development phases. Although the degree of involvement at different phases may differ by functional area, as shown in Table 3-1, each member should be a contributor, adding the valuable perspective of his or her particular functional area to the team's decisions and trade-offs. This early and continual involvement both heads off future problems and permits parallel, concurrent activities that can dramatically speed up the new product development process.

The new product development process as well as individual products should be candidates for continuous improvement.

Product Reviews

Another important part of the staged product development process is the establishment of major criteria or milestones that the product must meet in order to progress to the next phase of development. Because subsequent phases typically involve an increase in the level of resource committed by the firm, project gates are typically controlled by senior management.

It is important that these product reviews not become onerous. Management has the responsibility to make hard choices and decisions as products evolve from concept to launch. The balance of being involved and informed yet empowering the multifunctional team with responsibility and authority is a difficult one. Product reviews often require a

TABLE 3-1.   Common Product Development Team Involvement in Different Stages

Stage	Functional Area				
	Marketing	Engineering	Manufacturing	Sales	Service
Concept	X	X			
Development	X	X			
Design		X	X		
Manufacture/Launch	X		X	X	X

significant effort by the team and a chance to pull together. It is important to keep a high level of decision making and "empowerment" with the team. Flexibility and balance are the watchwords, for example:

When the new product is an upgrade or enhancement of an existing product (one of a product family), the team should be able to shorten the full new product development staged process.

Major new concepts involving significant technical or marketing risks and expenditures should come under increased scrutiny.

Management reviews should pay attention to the process as well as the product. Was QFD used to develop product specifications? Is market research on customer requirements adequate? Have major technical and competitive risks been addressed?

Senior management should take a balanced and flexible approach to the review and approval of new product development phase gates.

TYPICAL PROBLEMS

The project sponsor's emphasis is almost always unclear initially, and project personnel tend to assume their own biases in ranking the relative importance of each dimension. This can easily lead to a disastrous outcome, which can be avoided by adequate discussions between the new product development project manager and the members of the multifunctional team.

A second problem is that many project managers are asked—or compelled—to take on responsibility for a project that was planned by someone else. In some such cases, they do not know how to accomplish the Triple Constraint. In such a situation, the new project manager must take some time to study the project's scope, offer several options to his or her management (and, perhaps, the sponsor), and provide a recommendation on how to proceed.

A third major problem, mentioned earlier, is the myopic attention to the performance dimension by technical personnel. It can be overcome, or at least reduced, if the project manager clearly conveys the customer's emphasis and its rationale.

HIGHLIGHTS

- The Triple Constraint defines all projects.
- The Triple Constraint consists of performance specifications, a time schedule, and a money or labor hour budget.
- Obstacles that prevent satisfying the Triple Constraint are not mutually exclusive.
- Project specifics determine the relative importance of each dimension of the Triple Constraint.

- The new product development manager has two customers for the new product development effort: the external consumer and his or her own firm.

- The new product development project manager and the members of the multifunctional team must hold adequate and clear discussions to develop a common view of the Triple Constraint and what will be deferred for inclusion in later members of the new product family.

- Staged product development is an effective structure to bring products from concepts to the marketplace.

- New product development projects should normally be structured in phases (or stages) with thoughtful "go or stop" reviews at the end of each phase.

- Product reviews need to keep management informed, but decisions should stay with the multifunctional team.

- The staged new product development process needs to be customized to the individual company's unique requirements.

Notes and References

1. K. B. Clark and T. Fujimoto, "The Power of Product Integrity," *Harvard Business Review,* November–December 1990, pp. 107–119.
2. G. Hamel and G. K. Prahalad, "Corporate Imagination and Expeditionary Marketing," *Harvard Business Review,* July–August 1991, pp. 81–93.
3. S. C. Wheelwright and K. B. Clark, *Revolutionizing Product Development: Quantum Leaps in Speed, Efficiency, and Quality.* New York: Free Press (Macmillan), 1992, pp. 152–153.
4. R. Johnson, "Product Delivery Process," presentation 8 July 1992 at a meeting of the Los Angeles chapter of the Project Management Institute.
5. W. Cloake, "Concurrent Engineering in the Plotter Industry," presentation 10 July 1992 at a meeting of PDMA-WEST in Pomona, CA.
6. "Product Innovation Process—How to Turn Ideas into Commercial Products," commercial brochure published by Exxon Chemical Polymers Group, 1991.
7. T. MacAvoy, presentation 15 October 1992 at the sixteenth annual international conference of the Product Development and Management Association, Chicago.
8. G. Pinkerton, presentation 16 October 1992 at the sixteenth annual international conference of the Product Development and Management Association, Chicago.

CHAPTER 4

How to Start Successful New Product Projects

This chapter examines a few strategic issues that govern the initiation of successful new product development projects. Business planning—including product planning that goes far beyond the narrow single development project—is a central underpinning. Then we introduce the use of quality function deployment (QFD). Next we describe the role ISO 9000. Finally, we discuss checklists.

STRATEGIC ISSUES

Some organizations lack strategic discipline. Any new product project is good or desirable in this environment. But even very large companies have finite resources, and the more successful companies concentrate these on the few most attractive new product development projects. To do otherwise starves some (or all) key efforts for critical capabilities. The goal of this section is to provide you with some orientation to a few of the concerns that confront senior managers in considering which efforts to pursue or authorize.

Company strategy is concerned with its long-term direction and prosperity. This direction is perhaps best understood as a common vision shared throughout the firm. This common vision of the business, its objectives and values, can go a long way in uniting the overall "multifunctional team" that is the company.

Just like individuals, companies develop a personality or corporate culture. This culture is a set of values commonly held throughout the firm. They shape and determine how this entity will interact with its external environment, including customers, competitors, and legal, social, and ethical issues.

A company's vision of the firm is often captured in a short mission statement. Mission statements typically refer to technology, customer, or performance issues. A mission statement is important because of both content and length. Top management is forced to compress into a few sentences its communication on the firm's direction. Choices

The mission statement tells a lot about a company's values and culture.

must be made whether to include customers, products, shareholders, operations, or other aspects, as suggested in the following list of choices:

External forces	*Program management tools*	*Success criteria*
Company history	Business models	Product ROI or IRR
Existing customers	Project management	Launch date
Competition	software	Market share
Technical capacity	New product financial model	Performance compared
Manufacturing capacity	New product market model	to specification
Sales force	Team software	Development resources
Financial resources		and budget
Market position		
Support capacity		

The choices made are an important message in and of themselves.

As mentioned previously, TQM is having an important impact on the basic business philosophy of many companies and the way they formulate strategy. Although the TQM approach is designed to put higher quality products into customers' hands faster, the new ground rules contain subtle changes in the basic strategies to develop new products:

Customers. Long-term customer satisfaction becomes a major corporate objective. The requirement to define and lead customer needs becomes a guideline for product development.

Company. The culture of the company shifts from a hierarchical organization to a flat, team-based one. Indeed, new product development teams may lead the way in this change.

Competitors. Competitors are still to be bested in the battle for the customer; however, they also have input to the continuous improvement process. They become targets or "benchmarks" of performance for a particular product or business operation.

If adopted, TQM will have a substantial effect on the culture of the organization.

BUSINESS PLANNING

A critical step in the strategic planning process is the business implementation of the strategic direction. Unfortunately, there is often a tendency to put in place lofty strategic objectives without the hard commitments and resources necessary to implement them.

The long-range or strategic plan is usually broken down into divisional or functional business plans, as shown in Figure 4-1. A key objective of this process is to assure that all parts of the company are headed in a common direction. For example, in the case of a new product, marketing promotion plans should coincide with initial product availability from manufacturing and staffing from product support and sales training efforts.

The essence of business planning and control is to provide the mechanism to relate daily tasks, projects, and spending to longer term objectives. A firm needs to have the ability to know when progress is off track and the capacity to identify and execute corrective actions.

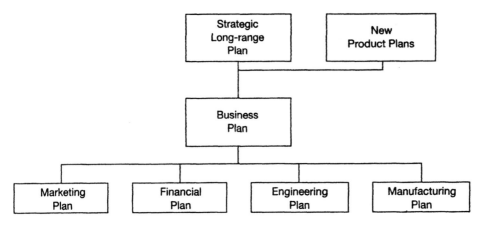

FIGURE 4-1.   The planning process.

Business planning addresses critical implementation elements, including details on products, budgets, development costs, product promotion, and advertising—all the activities necessary to make the strategy a reality. A key element of this process is the discipline to provide the all-important follow-through.

Hoshin Planning

Hoshin Planning or Hoshin Kanri, as it is sometimes called, is an approach that addresses the common problem of tying the control of daily operations to the realization of longer term, strategic objectives. Hoshin Planning uses simple forms that tie the process together.

Strategic objectives with specific short- and and long-term goals are listed. The value of the approach is in the structure and discipline provided. Each objective and goal has an identified "management owner" and progress milestones to be accomplished at specified times. Hoshin Planning can provide a roadmap for continuous improvement and disciplined implementation of strategic objectives and goals.

Business or implementation planning has two basic elements:

- The ongoing profitable management of operations based primarily on existing, currently available products.
- The development and integration of new products into these operations.

Historically, financial analysis and control have been at the core of this system. Although finances remain of paramount importance, many firms are expanding their vision to include market dimensions such as market share as well as quality dimensions such as customer satisfaction.

Business planning is concerned with the control and discipline to meet strategic objectives.

Business Planning Models

Strategic and business planning models are very helpful in providing a new perspective on the strategic plan and its implementation. These models show the impact of a wide range of variables on overall profitability and financial return. Among the more important variables that have a high impact on product profitability are the following:

- Time to market
- Market share
- Product quality

Computer-based models like PIMS (Profit Impact of Marketing Strategy) and Business Insight™ are also important tools for the new product development program manager. These tools capture the experience of both industry experts and other firms in similar businesses to provide a level of experience unavailable in most organizations. These models can be used to show the impact on product profitability of the trade-offs and priorities typically necessary during product development:

- Changes to product specification to reduce cost and accept a lower market share
- A decision to increase development funding or accept a slip in the product delivery date
- A decision to increase product reliability and increase product cost

Although these planning models won't make the decisions, they provide a useful checklist and reference point to help the team sort out and better understand the complicated, highly interrelated choices that must be made.

Business strategy models can provide both new insights and useful checklists.

Company Software and Models

Multifunctional product teams must also factor external product and market information into the internally developed models of the product. These internally developed models typically include one or more of the following:

- Project management schedule model
- Product financial model
- Product market model
- Product manufacturing model
- Product support model

Figure 2-2 described how these tools typically interact to help the decision-making process for the new product. Software packages that model the schedule, financial return, production, or staffing are simulation tools that help the new product development program manager in several ways:

- They help him or her understand complex relationships and interdependencies.
- They can be used to run sensitivity analyses of "what if" cases to test the importance of key assumptions or potential changes.
- They are consistent.
- They provide a checklist to ensure completeness.
- They provide for the documentation of assumptions.
- They aid communication both within the multifunctional project team and with senior management.

Another important quality tool, quality function deployment (QFD), which we describe in some detail later in this chapter, can also be used as a new product development planning tool. The visual, documentation capabilities of QFD make it particularly powerful in relating disparate considerations such as market segmentation, customer applications, and product specifications. QFD can also be used in conjunction with financial and marketing models to help in documenting and evaluating the effects of alternatives and product decisions.

Internal software and modeling tools are helpful for communication and decision making.

PRODUCT PLANNING

Typically, the responsibility for new product development rests with the engineering or R & D (research and development) department. In major firms, marketing plays a larger, occasionally dominant, role. In the ideal product development cycle, marketing identifies a bundle of customer-originated needs to which engineering responds with the appropriate product specification. In many situations, technical breakthroughs generate product improvements or innovations that are ultimately packaged and sold to the market. Usually, some mixture of this technology and market interaction sequence is at the germination of the new product development effort.

Japanese companies have employed the product development process as an offensive weapon. They typically use speed and their ability to bring a product to market quickly to gain a competitive advantage. One example is how Honda and Yamaha used this tactic in the motorcycle market:

In 1981, Yamaha took the offensive and introduced sixty new models. Honda introduced sixty-three.
Yamaha let it be known they would be number one in the motorcycle market within a year. Unfortunately, they "stepped on the tail of the tiger," and hard.
During the next eighteen months, Honda introduced eight-one new models and made 113 improvements. Yamaha was able to introduce only thirty-four new models and make 37 improvements.
By 1983, the war was over. Honda introduced 110 models to Yamaha's 23. Yamaha had to sell off assets to survive.[1]

An important strategic consideration in product development is often an evaluation of a firm's product "portfolio." This portfolio approach identifies balance and diversification as important considerations in the new product selection criteria. This approach is borrowed from a financial investment concept in which a diversified portfolio (stocks, bonds, gold, real estate, and so forth) is seen as a low-risk way to obtain a superior reward. An opposite high-risk strategy would be to put all the investment into one vehicle (such as gold), increasing the potential payoff but with a disproportionate increase in risk.

The portfolio approach to new product development suggests that a mix of products at various stages of their life cycle in different yet complimentary markets is a superior strategy. However, corporate and product planners must pay attention to the product mix across several dimensions:

Diversification in the product mix is a sound product development strategy to pursue.

Product life cycle. To balance cash flow, profitability, and corporate resources, the number of new starts and maturing "cash cow" products needs to be in balance.

Technical capabilities. The firm needs to go with existing technical capabilities; yet it must also continue to push into new product areas and complementary technologies.

Market capabilities. Firms enjoy a certain position with their existing and potential customers. Again the benefits of acquiring a new market position or opening up a new market segment must be weighed against competitive, market acceptance and investment risks.

Customer Applications

A key element that drives the new product specification is the end user customer application. It's often important to stress the *end user* customer for industrial products. Not only many industrial products may be handled by a multilevel distribution channel, but the product itself may be incorporated as part of a subassembly or system component for the ultimate end user product. This focus is not to diminish the importance of these intermediaries, but rather to highlight the importance of the ultimate consumer. The multifunctional team needs to understand how its product helps the end user customer do a better job and stay competitive. Without this perspective, product development efforts are at risk.

Connor Peripherals (San Jose, California) provides a good example of this approach. A rapidly growing supplier of disk drives for portable computers, Connor works hand in glove with its customers to stay ahead of the rapidly changing market. Because of the pace of this change, Connor recognizes that the only way to stay competitive is to become a member of their OEM customer's design team. Given its "sell-design-build" product development cycle, Connor must have a firm grasp of how its disk drive contributes to the acceptance of its customer's product.

Product Families

The product family concept fits well with both investment and market considerations. From an investment standpoint, having a family of products at different stages of their product life cycle provides a natural risk hedge. From a marketing standpoint, the application and customer satisfaction feedback from an initial product is important input to the development and targeting of follow-on products.

The product family concept is also consistent with the "kaizen" or continuous improvement philosophy mentioned in Chapter 1. The product development team needs to appreciate the power of customer feedback and experiences with existing products. Sometimes relatively simple, inexpensive product enhancements can open up new applications and address market segments not available to the initial product. Indeed, considerations of product modularity, compatibility, and on-site upgrade ability can be important considerations for product acceptance. Firms should view the new product development process as a continuous, never-ending process of serving a constantly changing market, competitive, and technical environment.

Product families are an attractive approach from both the development and market standpoint.

Market Research and Market Information

The effective use of market information is a critical element in the development of a successful new product. Most of the uncertainty in the product development effort is usually in the market and competitive arenas. Although it can never be eliminated, the appropriate use of market information can reduce the overall risk associated with the new product.

Market information has as its objective the identification, collection, and evaluation of existing and *anticipated* customer needs. Market information comes in many types and forms. We describe some of the general types of available market information.

Existing Customer Information

Although not strictly considered market research, the collection and analysis of existing information on customers should not be overlooked. From a certain perspective, the company has been carrying on a very large market research program—called sales. The history of successes and failures, applications, industry segments served, major customers, product use, and other data often provides an important backdrop and direct input into the market analysis effort for the new product. The availability and minimal cost associated with this type of market data are also advantageous.

Information on existing customers is important market data that should not be overlooked.

External Market Information

Market research information is taking on a new level of importance as the focus on customer needs and satisfaction increases. However, the possible choices and alternatives can make the selection of the appropriate type of market research a difficult task. As shown in Figure 4-2, market information exists in four general categories. Understanding the nature of these categories will facilitate acquiring the appropriate market data.

Quantitative.   This form of market research is the most common. It usually represents a large number of data points consistently defined over a large population. Design of the data collection as a statistical representation of populations is often required. Its direction is generally narrow and targeted in scope, but precise in its findings.

Type / Source	Quantitative (conjoint analysis)	Qualitative (focus groups)
Primary (field interviews)		
Secondary (public reports)		

FIGURE 4-2.   Market information types and selected examples.

Qualitative.   This type of market input tends to be more subjective in nature. Statistical representation is sacrificed to obtain a better understanding of the respondent's needs, motivations, or applications. Although the "data" are more palatable than findings from quantitative research, they are also indicative rather than conclusive in nature. Analysis of the data is not as straightforward, and the process often requires a high level of skill to administer and analyze correctly. Interviewing techniques include in-depth interviewing or structured focus group sessions.

Primary Research.   This type of research involves a custom-designed project to define requirements, identify respondents, acquire information, analyze data, and provide results. It is typically more expensive and time-consuming than secondary approaches but has the major advantage of being designed to meet the unique requirements of the specific project under consideration. Findings will be proprietary.

Secondary Research.   The proliferation of on-line data bases, compact disk read-only memory (CD ROM) technology, and other information sources has significantly improved the quality, cost, and availability of secondary market data. These types of data have the advantage of being relatively inexpensive and readily available. However, the challenge is in accurately defining both the applicability and quality of the data for your requirements. Because the information is public, use of secondary research can never provide proprietary intelligence—although there is always the possibility that your company may develop a unique insight. It is often a good idea to start off a larger primary market research effort by reviewing and consolidating the available secondary information.

Market data are available in many forms. It is important to recognize the availability, quality, and applicability of alternatives.

The Uses of Market Information

Although the uses of market information are frequently project specific, they tend to break down into two general categories:

1. Market and competitive assessment needs
2. Product-specific market requirements

Market assessment provides a realistic backdrop or environment for the new product.

- It collects information on customer needs, applications, and the motivations behind product purchase.
- It assesses the acceptance of technology from a user perspective.
- It describes the company's competitive position, its strengths and weaknesses in the market.

Product-specific information is frequently developed without an adequate assessment of the marketplace. That is the equivalent of building a magnificent structure on top of an inadequate foundation.

Because it is within the bounds of a defined product area, product market information tends to be relatively straightforward. Product market data have two primary uses:

1. *Market measurement.* This includes sizing the market opportunity for a new product, providing some indication of the overall market size and the share of the

available market the new product must capture. This analysis is usually a major part of determining the rationale behind the project revenue stream for the new product.

2. *Product design.* The mapping of product features and capabilities to customer needs is at the very core of the new product development process. We cover this process in more detail in the section on QFD, but it is important to make the product-to-customer connection a very strong one early in the new product development process. It is also important to recognize that matching customer needs and product specification is not adequate. Firms must try to *anticipate* customer needs in order to represent more closely the environment in which the new product will compete.

Market data are used both to describe the competitive environment and to size the market for the new product.

Phased Market Research

The new emphasis on customer satisfaction dictates that customer feedback serve as a continual guide during all phases of product development. The appropriate market information should be planned for each project stage to identify, monitor, and control the new product's continual focus on the customer.

Table 4-1 is a rough guide tying together the product development cycle and the type of market input likely to be required at different stages. The idea is to set in place a customer feedback plan that is proactive, not reactive, an approach to market information that not only assures a continuous stream of customer input but also ensures that the nature and timing of the information will be appropriate for the decisions and trade-offs required.

In the concept screening phase of a new product, information needs to be broad in order to describe potential unmet customer needs and competitive opportunities. Understanding the customer's problems, applications, unmet needs, as well as the direction of technology and competition are all important elements.

As the product enters the design specification setting stage, the major requirement becomes to size the market opportunity and get feedback on the attraction of proposed product function and features to the customer.

As the design specifications become firm, it becomes even more important to continue going back to the customer for input and guidance. In-depth understanding of potential applications and reactions to early test models is valuable to make the necessary product refinements prior to full production.

Launch is the "execution" of the new product development process. Advertising, sales and support training, and other activities must be monitored for corrective action

TABLE 4-1.   Types of Market Information That May Be Required at Different Times During the New Product Development Process

Product Phase	Market Information
Idea—concept	Secondary research, case studies, focus groups
Design—specification	Quantitative market sizing, product feature preference models
Prototype—initial production	Concept testing, beta test feedback
Product launch	Launch effectiveness feedback

in order to keep this critical process on track. At this point, it is often desirable to establish ongoing customer panels and customer satisfaction surveys to guarantee product success and set the stage for subsequent product family members.[2]

QUALITY FUNCTION DEPLOYMENT (QFD)

QFD fits well into the new product development process, permitting a proactive, anticipatory posture to setting the product specification rather than a reactive, quick fix, change approach. QFD requires an investment up front—an investment of time and resources to obtain and integrate the voice of the customer (VOC). QFD allows the VOC to drive the product specification and future phases of the development processes, including manufacturing, support, and so forth. Although the primary application of QFD has been the development and improvement of hardware products, the capabilities inherent in QFD are generic in nature. QFD is not unlike a spreadsheet capability that can be applied to a wide range of new product-related activities, including:

Different types of market information are needed at different stages of the new product development process. Careful planning of customer feedback can enhance the probability of the new product's success.

- Business planning
- Plant site location
- Design of market research projects
- Design of application and system software
- Development of service-based products

Wherever there is a need to capture detailed customer requirements and relate them to a process to obtain output, QFD has the potential to contribute.

QFD and Cross-Functional Interfaces

QFD is an important element in improving the new product development *process*. Figure 4-3 illustrates some of the elements in the top-level QFD chart. Matching WHAT the customer wants to HOW the product design will accomplish the task is critical. This step is at the core of multifunctional teamwork because it brings together marketing, engineering, manufacturing, and all the other involved functional groups. QFD is a powerful tool that can help you break down cross-functional disharmony and build multifunctional teamwork. It also involves an important *process* change in new product development. Rather than simply providing better market research or an improved design, QFD forces a focus on the process of how they interface. As shown in Table 4-2, QFD can be applied to either hardware or software new product development. The elements of a few charts are also listed in Table 4-2. Figure 4-4 illustrates where QFD fits into the new product development process.

QFD is a useful tool to help build multifunctional teamwork.

Because customer needs (WHATs) drive the QFD process, their importance quickly becomes obvious even to technical personnel who are more comfortable working at the implementation, back end of QFD. It is important that customer needs are actually developed from the customer and are not engineering's or marketing's guesses—or narrow viewpoint—of what the customer wants. In many cases, obtaining customer input involves some mix of the internal and external, quantitative and qualitative market information described previously. Although there is no single correct way to

QFD will not find a market opportunity, but its use can clarify it.

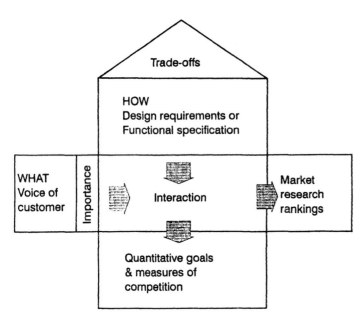

FIGURE 4-3.   Elements in the top-level quality function deployment chart.

proceed, Figure 4-5 illustrates a typical sequence. QFD does not, of itself, shape the market opportunity, but it is a very useful tool after you have defined the user's environment. The unique requirements of the new product development project should drive the market information acquisition process.

At various times, we have used a process of field interviews, mail surveys, telephone interviews, focus groups, affinity diagrams, and conjoint analyses to identify and make trade-offs of customer needs. Focus groups or in-depth individual interviews can be used to brainstorm product requirements, including specific examples to aid in under-

TABLE 4-2.   Common Names of and Outputs from Quality Function Deployment Charts for Hardware and Software New Product Development

| | Type of New Product Development Effort | | | |
| | Hardware | | Software | |
Chart Level	Chart	Output	Chart	Output
First	Product planning		Requirements analysis	
		Design requirements		Functional specification
Second	Part deployment		Architectural design	
		Part characteristics		High-level design
Third	Process planning		Technology assessment	
		Manufacturing operations		Methods and tools procedures
Fourth	Production planning		Implementation planning	
		Production requirements		Resource plan

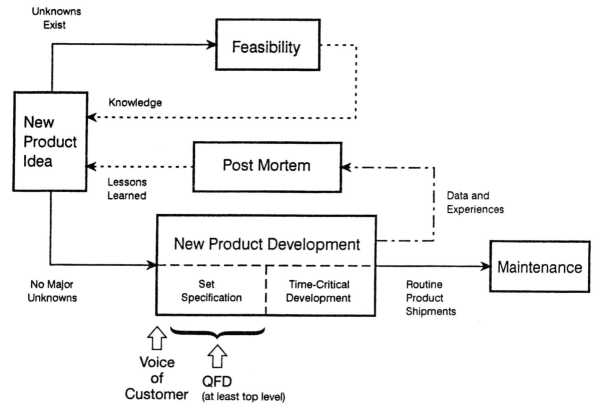

FIGURE 4-4. Quality function deployment is applied to help set the specification in the new product development process.

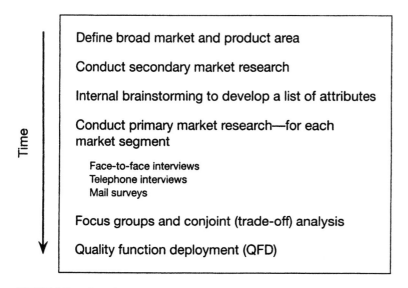

FIGURE 4-5. Typical time sequence for capturing the "voice of the customer."

standing. Analysis and coding of the information obtained from the interviews is a vital step. It is important that this be a cooperative exercise among team members because the output (the WHATs) will drive the QFD process.

It is critical not to get prematurely wrapped up in the market research process. A sizable project to gather input from hundreds of respondents across several market segments is not appropriate for an initial QFD effort. It is probably better to start off small with existing market data and fewer interviews to get an initial familiarity with the requirements of the process. Appendix E contains additional details on QFD and describes two very different projects to illustrate the potential use for this powerful tool.

QFD and Benchmarking

Benchmarking is a process for measuring performance against some standard or reference. It may address a range of market-related, external issues, including customer service, sales effectiveness, and new product acceptance. Benchmark projects also focus on internal operations where the objective is to improve activities such as sales order entry, product testing, or billing procedures. Benchmarking has imitation at its core. The process simply identifies the world class leaders in a particular product or functional area, understands the elements of their success, and incorporates the approach into the targeted business operation. An important element of benchmarking, as with all quality-based tools, is the focus on the customer as the measure of value. These are the four key steps in the benchmarking process:

1. *Planning.* Define the area to be benchmarked. Identify the best in class competitors. Plan data collection and analysis techniques.
2. *Analysis.* Measure and evaluate the differences between the competition and your performance.
3. *Integration.* Integrate the goals and objectives into the new product development or other process.
4. *Action.* Define specific plans. Develop feedback to track progress. Provide for recalibration of the benchmark.

A helpful benchmark effort often used as part of the new product development process is to look at the product area in which the new product will compete. After identifying the leading products in the class, it is useful to obtain a customer-driven appraisal of their performance. As an example, it often can be valuable to complete a table such as Table 4-3 for existing market leaders in the new product area. The following are some of the benefits your new product process might gain from such a table:

1. It drives the effort with customer-generated input.
2. It indicates both the customer's importance and satisfaction level to help target improvements for the new product or a subsequent model.

TABLE 4-3.   Table for Recording Competitive New Product Performance Information

	Product Performance	Marketing Sales	Customer Service	Administrative Support
Customer importance				
Customer satisfaction				

3. It helps expand the concept of the product beyond hardware to include service, sales, and support.
4. It is a natural lead-in to the QFD process.

Before initiating a benchmark product, you must consider three important points:

1. In order for benchmarking to work, the organization should be willing, even eager, to implement the changes that will inevitably surface.
2. Companies are different. It's important to custom-fit the benchmark results to *your* firm.
3. Benchmarks are moving targets. Be prepared for continuous improvement.

QFD and benchmarking are complementary procedures. Benchmarking can help identify areas that are high payoff candidates for improvement. QFD is the tool to help in implementing specific improvements revealed as needed by the benchmarking process.

In fact, the QFD preplanning matrix (see Appendix E) is an ideal form to gather and structure both the satisfaction and importance assigned by the customer in the benchmarking process. Given the detailed nature of the QFD process, this preplanning preparation ensures that only high-value (from the customer's standpoint) aspects of the new product will undergo the scrutiny of the QFD process.

ISO 9000

The European Community (EC) has taken an important step to ensure that uniform quality standards exist between its members and anyone doing business with them.[3] The International Standards Organization (ISO), based in Geneva, Switzerland, published the ISO 9000 Series Standard in 1987. The objective was to ensure that suppliers of goods and services met certain uniform minimum quality requirements.

The ISO standards currently consist of five sections:

1. ISO 9000. An overall guideline for selection of the appropriate standard
2. ISO 9001. A standard describing supplier requirements to design and provide the product
3. ISO 9002. A standard describing the requirements for a supplier to show the capacity to control the process for accepting products as meeting specification
4. ISO 9003. A standard outlining how a supplier can demonstrate the capability to dispose properly of nonconforming products
5. ISO 9004. A standard describing the approach to develop and implement a quality management system[4]

These international standards have also been adopted and interpreted by individual countries. For example, the U.S. standard is known as "Q90"; it is "BS 5750" in the United Kingdom. Some difficulties have been noted as each country interprets the standard to its individual language and culture.

The implications of ISO 9000 are important for the new product development professional, particularly the manufacturing team member. ISO 9000 can be viewed as a basic level of quality conformity that must be met in order to do business with the EC.

There is a formal audit and qualifying process in place. Certification is done on a manufacturing plant basis. Depending on the size and complexity of the company, the process can involve considerable time and expense.

Although the effort for certification is a significant one, many firms view the process as a cost of doing business in Europe. Beckman, a leader in the quality area, was among the first U.S.-based companies in the life sciences to achieve certification. Plants in Carlsbad, Fullerton, and Brea, California, as well as Galway, Ireland, were recently qualified. However, the benefits went beyond certification. Lloyd Bostwick, manager of Reliability Engineering, noted, "But then, this is the first time every element of our operations has been brought together and addressed by a single document. There is certainly a lot more continuity now."[5]

CHECKLISTS

Checklists are designed to help assure that nothing that will have to be dealt with during the course of the new product development effort has been forgotten or omitted from consideration. A checklist might contain such items as shown in Appendix C. This checklist is not exhaustive and may not contain the most significant or most important items for your new product development effort. It is meant to suggest the kinds of items that might in appear in a checklist.

The best way to develop a checklist is to create your own over a period of years. One way is simply to enter items on file cards or a word processing file alphabetically, by time phase, or by some other logical method. Having developed a checklist from your own experience, you will perform better on future new product development projects because you will be unlikely to forget items that are potentially significant. Thus, you will consider their impact on a new product development effort in an early planning phase;

Take advantage of experience.

they will not emerge as unexpected requirements during the performance of the time-critical development work.

TYPICAL PROBLEMS

The use of a staged new product development procedure runs the risk of impeding simultaneity to shorten time to market. Concurrent engineering, for instance, will become impractical if the product design work is contained in a phase different than the process design work. Thus, you must establish your stages to assure that multifunctional teams work in all phases and that activities are overlapped to the maximum extent practical, consistent with assuring end-of-phase reviews to reduce risk.

Another problem is deciding when to abandon a new product development project. Staged approach is helpful in requiring periodic scrutiny of each effort, but the new product development project manager and the multifunctional team may be reluctant to recommend cancelation even if they believe they can no longer achieve the goals. The corporation must support a climate in which stopping a new product development effort is not seen as a failure but rather a case of having become smarter.

A third problem is deciding to suspend a new product development effort that is meeting all its milestones. This sometimes must be done to assure that adequate resources are applied to a higher priority effort.

HIGHLIGHTS

- Companies must concentrate their limited resources on new product development efforts that support and are consistent with the company's strategy.

- Products must not be planned in isolation.

- Families of new products offer many advantages.

- Quality function deployment is a powerful planning tool.

- A checklist may help you avoid overlooking required work.

- A firm's mission statement is an important common vision or shared value within the company.

- TQM requires that customer satisfaction play a major role in a firm's strategy and new product development efforts.

- Planning and financial models can be of major assistance in documenting, monitoring, and evaluating the new product development effort.

- Market acceptance is usually the most uncertain but unfortunately the most important element in the financial success of the new product. Customer input is vital.

- QFD is a system that integrates the voice of the customer into a product's design and production.

Notes and References

1. R. B. Kennard, "From Experience: Japanese Development Process," *Journal of Product Innovation Management,* September 1991, pp. 184–188.
2. Developed from a V. Vaccarelli presentation, 1 November 1990 at the fourteenth international conference of the Product Development and Management Association, Marina del Rey, California.
3. See, for example, W. Winchell, *Continuous Quality Improvement: A Manufacturing Professional's Guide.* Dearborn, MI: Society of Manufacturing Engineers, 1991, pp. B4-B7; and T. E. Benson, "Quality Goes International," *Industry Week,* 19 August 1991, pp. 54-57.
4. J. Holusha, "Global Yardsticks Are Set to Measure 'Quality.'" *New York Times,* 23 December 1992, p. C6z.
5. *Beckman Life,* First Quarter 1992, p. 4.

PART 3

Planning a New Product Project

CHAPTER 5

Why and How to Plan a New Product Project

The planning activity for the management of a project is crucial. Plans are the simulation of a project, comprising the written description of how the Triple Constraint will be satisfied. Therefore, project plans are really three plans: one for the performance dimension (the work breakdown structure), one for the schedule dimension (preferably a network diagram but occasionally a milestone listing or bar chart), and one for the cost dimension (a financial estimate). This chapter covers plans in general, reiterates the need for plans, describes how these three kinds of plans are made, and reviews several planning issues.

PLANNING

In broadest generality, plans depend on three factors:

1. Knowing where you are now (or will be when whatever is being planned for will start)
2. Knowing where you want to get
3. Defining which way you will get from where you are to where you want to be

These factors are illustrated in Figure 5-1. The old saying, "When you don't know where you want to go, any road will get you there," is true; you can have a plan only if you have a destination in mind.

Most new product development is originated because of the corporation's long-range plan. Thus, plans are frequently hierarchical, with short-range plans established within the context of long-term plans. For instance, project task plans are components of the overall project plan. As mentioned in Chapter 2, task plans are components of project plans and so on. In addition, planning is iterative, so sometimes new product project plans must be revised when other plans change. For example, when the long-range plan covers five or ten years, changes obviously occur, priorities must be altered, and projects are added or canceled in response to the dynamic environment.

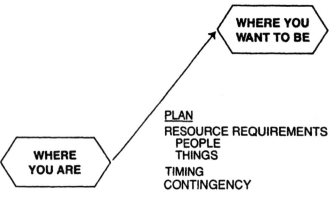

FIGURE 5-1.   Planning.

THE NEED FOR PLANS

Plans aid coordination and communication, provide a basis for control, are sometimes required as part of a corporate funding request, and help avoid problems.

Coordination and Communication

Most projects involve more than one person. Typically, a technical expert is asked to perform in the area of his or her expertise. For instance, an expert on electronic circuit design works on the electronic circuit design task, not on the optical design task for an electro-optical product. The project plan is a way to inform everyone on the project what is expected of him and her and what others will be doing. Plans are a vehicle to delegate portions of the Triple Constraint down to the lowest (task or subtask) reporting level. If the people responsible for these tasks also participate in making the plans, they will have an added impetus to adhere to them. Thus, there is a Golden Rule for planning:

Let others plan their work.

Get the persons who will do the work to plan the work.

- They should know more about it than anyone else.
- It's *their* task, not *yours*.

Your project plans matter. Even if your project can be performed in your office, other people in the organization (for instance, your boss) will want to know where your project is headed, what you are doing, and for how long you will be doing it. Thus, project plans constitute an important communication and coordination document and may motivate people to perform better. The relationship of key new product development

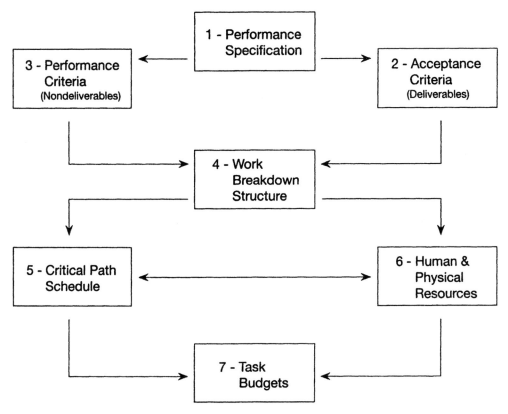

FIGURE 5-2.   The relationship of new product project plan elements.

project plan elements is shown in Figure 5-2, and the following discussion is keyed to that illustration.

1. The first requirement is a complete product performance specification. This should include items such as those we enumerated as examples of product attributes in the outline list toward the end of Chapter 2. This may be considered the statement of work (SOW) for a new product project or program. Although this may be prepared by an individual in any functional department or by a team within any single department, it is best prepared by a multifunctional project team employing market-based quality function deployment (QFD) or a similar specification-setting process.
2. Specific, measurable acceptance criteria for all deliverables must be explicitly stated. These are based on the performance specification and provide an unambiguous target for the development team to aim at. These criteria may clarify that the multifunctional new product development team must also develop test equipment for use during the development effort or certify some aspect of the new product's performance before it can be shipped.
3. There may also be performance criteria or standards that must be complied with

that are not deliverable as such. For instance, your company may wish to comply with ISO 9000, may have environmental, safety, and health (E,S,& H) requirements to be satisfied, may require that special procurement practices be observed (for example, use standard parts or specific vendors), or may presume that corporate styling standards will be maintained.

4. The work breakdown structure (WBS) is derived from these first three items. The WBS (discussed in Chapter 6) identifies every work package (or activity or task) that must be completed by the multifunctional project team to satisfy the performance specification and other criteria. Most of the WBS will derive directly from the performance specification itself. However, as we mentioned, the acceptance criteria may stipulate certain test conditions. Thus, additional WBS elements may be required to prepare test facilities or other means to verify completion. Similarly, the performance criteria for nondeliverables may require the inclusion of further WBS elements. New product development project managers need not be intimately familiar with every WBS element, but they must have confidence that at least one other member of the multifunctional project team is fully competent to handle each WBS task. Alternatively, the team as a whole must have confidence that there is a way to complete every WBS task in a timely way.

5. Every element in the WBS must be entered into a critical path schedule network (discussed in Chapter 7). Milestones and bar (Gantt) chart schedules may be derived from this.

6. A resource plan is required to accomplish the WBS. It includes both human and physical resources, as was otherwise illustrated in Figure 1-5. The schedule depends on the resources that will be committed, so these two plan elements obviously interact. For instance, a senior person with relevant experience should be able to carry out a specific task more quickly than an inexperienced junior person. Similarly, if certain resources are overcommitted when it is desired the work will be done, the schedule will be unrealistic unless it is adjusted. Obviously, the resources available for a project depend on the resources dedicated to other work and relative priorities. Consequently, a new product development project can be delayed by other corporate activities.

7. Specific budgets for each task can be created, based on the resources that will be dedicated to the task and the specific time at which it is scheduled. If appropriate, a benefit-to-cost calculation or a discounted cash flow analysis can then be completed to determine if the development effort is justified.

As we have stated previously, the multifunctional project team must be comfortable that each task (4) will be satisfactorily completed on schedule (5) by the available resources (6). If so, that will determine the development budget (7), which is one ingredient in the new product's financial plan. Although the development expense is generally the least important financial aspect—getting the new product to market fast while achieving all the performance goals is normally what drives the financial return—it is not unimportant. Figure 5-3 illustrates where these project plan elements fit into the new product development process. It also should be clear that the Triple Constraint definition depends on the resources that senior management will truly dedicate to the new product development project, as shown in Figure 5-4.

The multifunctional new product development team must believe the project's Triple Constraint is achievable.

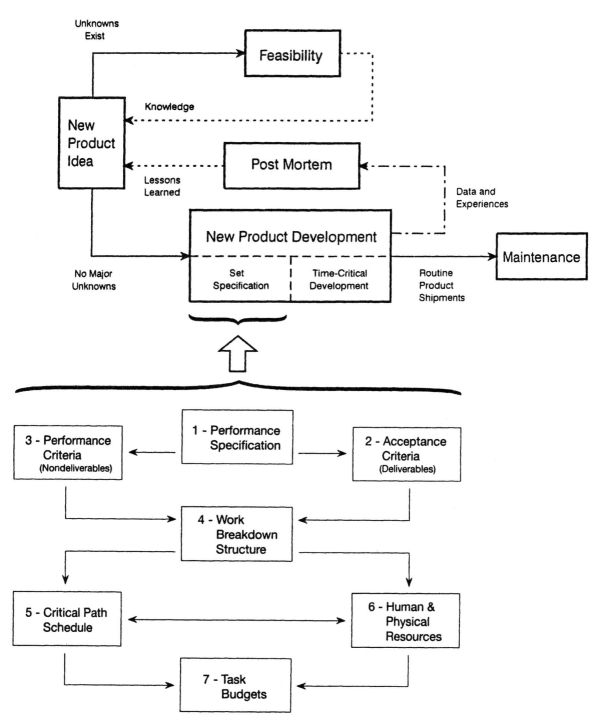

FIGURE 5-3.   The relationship of the new product project plan elements and the new product development process.

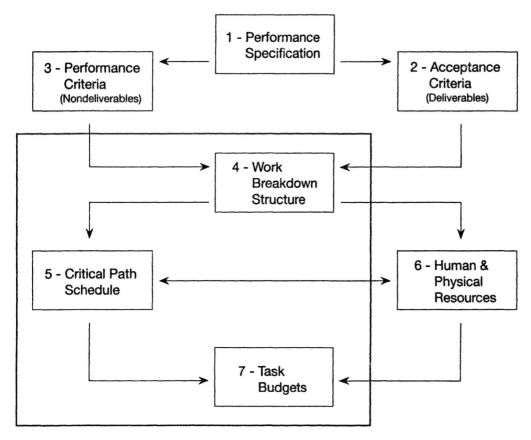

FIGURE 5-4.   The Triple Constraint depends on the resources that will be committed to the new product development effort.

Basis for Monitoring

Plans are also the basis of your project monitoring activity. It is a characteristic of projects that they do not go in accordance with plan. What you do not know when you start is where and how your project will go off plan. Deviations from plan, detected by monitoring progress, constitute your early warning signal during project performance that there are problems to be resolved. That should cause replanning to occur. Part 5 of this book discusses how variance information can be used to help manage the project.

Products do not follow their plan.

Plans are a detailed description, formulated before the project is carried out, for accomplishing its various aspects. Deviations may indicate that the project will not reach its intended destination.

There are plans for all three dimensions of the Triple Constraint.

Requirement Satisfaction

Plans are sometimes created to satisfy requirements imposed by others, perhaps your boss. In such a situation, plans are often created under duress rather than because they are perceived to be valuable, even essential, in achieving project objectives.

Plans so created are frequently not followed. All too often they are generated and then discarded because they were prepared only to meet the requirement to prepare a plan. When there is such a requirement and the plans are prepared slavishly rather than thoughtfully, it is a waste of time for the preparer and the reader.

Problem Avoidance

New product project management is sometimes a race with disaster. This is frequently the case for less experienced project managers. All too often the last crisis has scarcely been resolved before the current one begins, and then the project manager is too busy to anticipate and try to head off the next one. A good plan helps you avoid problems during project performance (but plans cannot prevent problems).

Planning is crucial; a good plan is essential.

PLANNING ISSUES

Uncertainty and Risk

Plans relate to future events. That is, your plans are a simulation of how things will occur in the future. There are necessarily uncertainties about the future, some of which may be somewhat predictable and thus partially controllable, but many of which are unpredictable.

Your objective is to be as confident as possible about those things that are predictable. You would also like to be fully confident that you will successfully achieve the Triple Constraint. But there will always be uncertainties, unanticipated tasks and outcomes, unexpected options, and unfortunate mistakes; so total confidence in complete success is naive. The best you can do is work toward all the objectives you can clearly perceive and recognize that there will be many totally unexpected developments that will challenge you and your team.

You can reduce (but not eliminate) these predictable uncertainties by using checklists, thoroughly discussing the plans with experts, and involving your entire team. Appendix C contains a general checklist for a new product development project. Nevertheless, uncertainties will remain because there are always unpredictable factors when you are doing something new. Allowance for these unknowns can be made by inserting contingency in your plans, but the unknowns cannot be eliminated. For instance, thorough plans cannot eliminate cost changes due to currency rate fluctuations. Plans can be no better than your present understanding. If you have done something similar before, you can plan it better than if it is entirely new to you and your team. For instance, previous experience with an organic polymer formulation is not terribly helpful for planning a new software product.

Use checklists to reduce uncertainties.

Assumptions, such as which people will be able to work on your project, are involved in planning. The plan for a new analytic instrument product looks a lot different when a senior mechanical engineer will do the work than when a junior electrical or chemical engineer will. Because assumptions are involved in your planning, it is important to include contingencies, which we discuss in Chapter 9. Good plans are quantitative rather than qualitative and as precise as possible.

A Choice Between Options

In preparing plans, as in carrying out project work, you are frequently confronted with options. These choices may include program management options, product quality standards, extent of subcontracting to be undertaken, and so on. Your plan may be considered the record of your choices between these options and will normally depend on how much risk you are willing to take or how much contingency allowance is included in your plans.

Project participants will frequently present a plan that seems absurd to you. It may in fact be absurd. But perhaps the person who prepared it is simply emphasizing activities that you are not stressing.

Heed others' plans and suggestions.

A common project activity, ordering required materials, illustrates this problem. Sensible choices are to order these materials as early as possible (to be certain they are available when required) or as late as possible (to reduce the possibility of having to change selection or to help your organization's cash flow). It is important to discuss the perceptions of everyone involved in the undertaking.

Hazards

There are innumerable hazards in preparing project plans. In an attempt to gain time in the early phases of a project or because you are addicted to your own ideas, you may tend to do much of the planning yourself. You should avoid doing so for the same reason you do not like to be told to carry out somebody else's plan: It is demotivating. In fact, it is important to involve the people who will actually be doing the work so they plan as much of their work as possible. Again, this is the Golden Rule.

Planning their own work can motivate people.

In addition, poor planning frequently occurs. Other than sheer laziness, the basis of almost all poor planning is a misunderstanding of the Triple Constraint point. Taking the time to create plans allows you to identify your perception of the Triple Constraint point and shows if and where it differs from somebody else's.

Occasionally, a tool commonly called a planning matrix is used. Figure 5-5 provides a partial illustration of such a matrix. It lists activities to be carried out along one side of a piece of paper and designates involved personnel along the perpendicular side. Where these rows and columns intersect at an "x," the designated personnel are involved in the designated activity. This kind of document may be helpful to some managers, but it is a misnomer to call it a planning matrix rather than an involvement matrix. To put it another way, a planning matrix may be a helpful document, but it is not a plan.

Currentness

Once you have decided to plan your project and have issued the plans, people should take them seriously. They can do so only if they know the plans are current. Therefore, it is very important to know who has copies of them. When you revise plans, be absolutely certain to provide revisions to all people who have copies of previous plans. When you do this conscientiously, everyone involved in your project will know that you take planning seriously. They will know the plans they have reliably indicate the project intention. You can increase others' assurance by dating all planning documents, and revisions *must* have a revision serial number and date.

Keep everyone current on revisions.

Task or Activity \ Function	Marketing	Engineering	Quality Assurance	
Market Research	X	X		
Product Specification	X	X	X	
Preliminary Design		X		

FIGURE 5-5.   Partial example of a planning matrix, showing the involvement of groups in project activities.

USING MICROCOMPUTER SOFTWARE

Project management software of any sort allows you to enter data so that the work breakdown structure, schedule, and budget are all consistent. Specific software packages allow you to display these data in varied formats.

The convenience, economy, and proliferation of microcomputer-based project management software encourage frequent minor plan revisions. In fact, there is a tendency for occasional major overhauls of the plan. This exacerbates the potential for different people working on the project to have different versions of the plan. Obviously, if the plan is maintained in a local area network (LAN) and the data base has file lockout, you can be assured that everyone has the same version. However, you cannot be sure that key persons have looked at their portions of the plan recently and noted any changes that affect their work.

Spend no more time planning than you would spend correcting problems resulting from having no plan.

TYPICAL PROBLEMS

There are four pervasive problems with planning. First, taking enough time to plan is costly. There is an old saying about this: "We don't have enough time to plan now; but we'll have lots of time to fix it up later." In fact, a little inexpensive planning early usually avoids a lot of very costly fixing later. It is axiomatic that software bugs are very expensive to find and remove when a programming project is far advanced. This is analogous to testing engineering breadboards and prototype models—a lot of early testing is much cheaper than trying to remove produc-

(continued)

tion problems later. It is difficult to decide how much planning is appropriate, but the inexperienced project manager usually does far too little.

Second, plans are frequently ignored because they are perceived as an irrelevant requirement of management. The solution is obvious: Write meaningful plans you intend to follow and keep current, and be sure everyone understands you have done so.

Third, a separate plan is required for each of the three dimensions of the Triple Constraint. These three plans must be integrated and consistent, and must not—as is frequently the case—be prepared by separate groups of specialists. Project management software can help assure consistency.

Fourth, some project managers are unwilling to prepare plans, formal or informal. These people should be reassigned to some other role because a project plan is an absolute requirement of success.

HIGHLIGHTS

- Plans are used for the following purposes: simulate how the project will be carried out; write the proposal; negotiate the contract; coordinate and communicate; increase motivation of participants; control the project; satisfy requirements; avoid problems; and record the choice between options.

- Plans delegate portions of the Triple Constraint to the lowest reporting level.

- Plans help keep projects on course.

- If formulated only to meet someone's requirements for them, plans are virtually useless.

- Everyone involved must receive every plan revision.

- Never spend more time on a plan than would be required to correct problems resulting from a lack of a plan.

- Plans reflect the balance between risk and contingency for both controllable and uncontrollable future events.

CHAPTER 6

The Work Breakdown Structure

The goal of a performance dimension plan is to be sure that everything required to satisfy the entire performance specification is done. This chapter deals with planning for the performance dimension of the Triple Constraint. The statement of work is a useful aid, but the principal tool discussed is the work breakdown structure.

STATEMENT OF WORK

The statement of work (SOW) enumerates what the new product development team will do and deliver. The SOW may be contained in a memo or work authorization rather than in an elaborate document or request for corporate funding, but it should still contain a specific, measurable, and attainable goal. The SOW must always contain a list of all deliverables and should be accompanied by a project schedule and budget to be meaningful. Thus, a plan for the performance dimension of the Triple Constraint is primarily a listing of every activity that must be performed and every result that must be obtained. The SOW frequently contains explicit acceptance criteria and test specifications.

WORK BREAKDOWN STRUCTURE

Purpose

The work breakdown structure (WBS) is a convenient method for dividing a project into small work packages, tasks, or activities. A WBS reduces the likelihood of something dropping through a crack. To put it another way, a WBS is intended to assure that all the required project activities are logically identified and related.

Use a WBS to subdivide projects into tasks.

    Figure 6-1 illustrates two possible WBSs for a new product, namely, the proverbial better mousetrap. There is no magic formula for constructing a WBS. The WBS may describe all the tasks that the multifunctional project team has to carry out prior to the new product's launch or merely the tasks to complete the current phase. If it is the latter

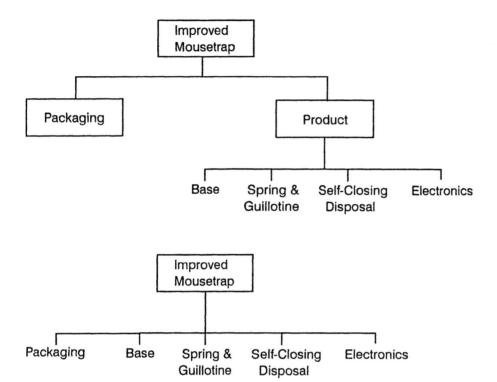

FIGURE 6-1.   Two illustrative work breakdown structures for a new product development project.

situation, with undefined work required in later phases, the desired shipment schedule is likely to be in jeopardy. The upper part of Figure 6-1 shows two levels of detail, but there is no standard number of levels to use. In general, probably at least three or four should be shown, but it might sometimes be appropriate to show five or ten or even more. The breakdown might occur using earlier or later activities, particular organizational involvements, or almost anything that makes reasonable sense.

In general, it is best to structure the WBS on tangible, deliverable items, both software and hardware. If your staged new product development process has clearly defined requirements for the end of each phase, then each of these will be a WBS task to be scheduled during the appropriate phase.

The more work packages you have in your project, the smaller and cheaper each work package becomes. However, the more work packages you have, the more money and time are spent in arranging for these to be properly interfaced with each other and managed. As we discuss more fully in Chapter 16, small WBS tasks with short durations improve the precision of project status monitoring. Conversely, if you have only one work package, there is no interfacing cost, but the task itself is large and expensive. Therefore, there is a happy midpoint that must be found by experience. In general, you should break your project into work packages sufficiently small so that each is understandable.

TASK AUTHORIZATION	PAGE OF
TITLE	

PROJECT NO.	TASK NO.	DATE ISSUED

STATEMENT OF WORK:

APPLICABLE DOCUMENTS:

SCHEDULE

START DATE: COMPLETION DATE:

COST:

ORIGINATED BY: DATE:	ACCEPTED BY: DATE:
APPROVED BY: DATE:	APPROVED BY: DATE:
APPROVED BY: DATE:	APPROVED BY: DATE:

FIGURE 6-2.   Task authorization form.

The more closely each WBS task conforms to prior experience, the more realistic and accurate your plan's schedule and cost estimate become. Another consideration in deciding how large a single WBS task should be is whether it will be the responsibility of a senior or junior person and their relevant experience.

The WBS defines the work packages and will be tied to attendant schedules and budgets for the work performers. Thus, it is desirable for the lowest level packages to correspond to small work increments and short time periods. It is often helpful to indicate who the task leader is by putting his or her name in the WBS box for the task to clarify organizational responsibility. It is also possible to insert the WBS number and schedule or budget information.

In addition to using the WBS for planning, it is sometimes helpful to use task authorization forms, such as in Figure 6-2, to clarify the statement of work for each task. If the task schedule or budget is a constraint, the task authorization form should have a

The WBS must be tied to time and money plans.

blank for this. Note that the task authorization form has a block where the task leader accepts the task, which assures compliance with the Golden Rule discussed in Chapter 5.

Helpful Hints

In preparing a WBS, do not forget required tasks such as analyses or trade-off studies that must be done. Also remember to include reports, reviews, and coordination activities. In fact, displaying them on a WBS is a good way to highlight that they are necessary and that resources must be devoted to them.

Fortunately, when a WBS is prepared, it tends to stress hardware integration activities. That is, junctions on the WBS frequently imply a hardware assembly or a test activity that must occur when these things are joined. Thus, a WBS again is useful for identifying an activity to which resources must be devoted.

If you can afford the time, it is desirable to have another person make a WBS for your project, independent of yours, at least down to the third or fourth level. This will take only an hour or so and will highlight any discrepancies or oversights. This approach may suggest a more effective way to organize the required work. You will have to repay the favor on later projects, but that should help your organization by reducing problems on projects. In fact, some organizations require that two or more people independently prepare a WBS for a given customer-sponsored project before it can be approved.

After the initial WBS has been made, schedule planning can commence. The schedule planning may identify further items to add to the WBS. Although less likely, the same may occur as cost planning is done. The WBS is then revised to include these work packages so that everything on the WBS is finally tied to scheduled work packages and budgets and vice versa.

Others can help assure that your WBS is complete.

MICROCOMPUTER SOFTWARE

Microcomputer-based project management software packages can be helpful in structuring the WBS for your new product development project. If the WBS tasks are entered in the hierarchical outline you and the multifunctional project team find convenient, most of the commercially available software packages will print a WBS chart (similar to Figure 6-1). Many of the other output options will also list the tasks in this outline form, which can be helpful. Appendix F provides a few examples.

TYPICAL PROBLEMS

> Vagueness in the SOW is a crucial problem in planning the performance dimension. For instance, the SOW may state that "appropriate tests will be performed." Who decides, and when, what is appropriate? The solution is to write a specific and detailed SOW.
> Another problem is the blind copying of a prior project's WBS for a new project.

When this occurs, we have not a WBS, but a waste of everybody's time—the people who prepare the WBS and the people who must read it. A project's WBS should be prepared thoughtfully, not by rote, to increase the odds of project success.

A third problem occurs when you prepare a WBS for any project and fail to revise it as subsequent schedule and cost planning reveal other work packages that are required.

HIGHLIGHTS

- A work breakdown structure identifies all work packages required on a project.

- A co-worker's independently produced WBS for your project may identify omissions on your WBS.

CHAPTER 7

Scheduling Tools and Time and Cost Estimating

This chapter deals with the second and third dimensions of the Triple Constraint. The plan for the schedule dimension orders activities so you can identify the logical relationship between them. In general, there are three approaches to scheduling: bar charts, milestones, and network diagrams. We discuss each in this chapter, stressing network diagram usage. This chapter also covers the important topic of time estimating, which must be done regardless of which scheduling tool is chosen. The plan for the cost dimension determines the development expense, and this chapter tells you how to estimate costs.

OVERVIEW OF SCHEDULING METHODS

Table 7-1 provides an overview of the main scheduling methods. Bar charts, such as shown in Figure 7-1, portray the time schedule of activities or tasks, and milestone charts portray the schedule of selected key events. Network diagrams portray activities, events, or both and explicitly depict their interdependency with predecessors and successors.

Time-based activity-on-arrow (TBAOA) schedules, which we favor, portray both interdependency and time sequence, as shown in Figure 7-2. (Table 7-2 describes further the

TABLE 7-1.   Comparison of Scheduling Approaches

Dependency Picture	Linear Time Scale	
	No	Yes
No	Lists of tasks or milestones	Bar (Gantt) charts or milestone charts
Yes	Network diagrams— Event-in-node (PERT) Activity-in-node (PDM) Activity-on-arrow (ADM) Hybrids with anything in node or on arrow	Time-based activity-on-arrow (TBAOA)

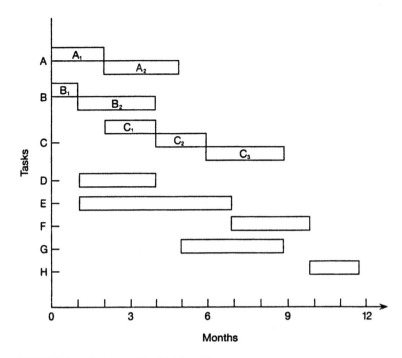

FIGURE 7-1. Bar chart with subtask breakdown.

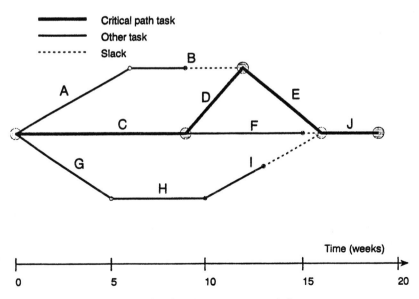

FIGURE 7-2. Illustrative time-based activity-on-arrow network diagram.

TABLE 7-2.   Illustrative Tasks and Durations for Figure 7-2

Task		Duration (weeks)
A	Formulate critical market research questions	6
B	Recruit focus groups	3
C	Preliminary product design	9
D	Build focus group demonstrators	3
E	Conduct focus groups	4
F	Finish design drawings	6
G	Survey new manufacturing technology	5
H	Analyze viable manufacturing equipment options	5
I	Study factory rearrangement requirements	3
J	Initial QFD	3

tasks illustrated in Figure 7-2. If all the arrows in Figure 7-2 had been drawn horizontally with task connections as vertical lines indicating predecessor-successor connections, the TBAOA figure could be thought of as a bar chart with task linkages added.

You can use the critical path method (CPM) with any of these formats by accenting or emphasizing the tasks (or activities) and events that must be completed on schedule to assure that the entire project is not late. A TBAOA schedule can help integrate the multifunctional project team's efforts because it clarifies who must do what by when so that the next task(s) can commence.

The time-based activity-on-arrow form of network diagram is extremely valuable to coordinate the work of the multifunctional project team and promote cooperation.

Using a discounted cash flow financial analysis (see Appendix D), you can easily calculate the benefit-to-cost ratio for a "crash" schedule acceleration, the impact of being late, and other schedule consequences. Similarly, you can assess the resource alternatives (for instance, senior people rather than junior, automated processes rather than manual, and so on). In any analysis of this sort, the development schedule and cost depend on the resources that realistically will be dedicated to the new product project. Wishful thinking about these parameters is "garbage in" and produces useless "garbage out" (so-called "GIGO") forecasts for the new product's prospective return on investment.

BAR CHARTS

Bar charts (often called Gantt charts) are frequently used for scheduling. Figure 7-3 is a bar chart. The project is divided into five activities with a planned duration of twelve months. When the bar chart was constructed, five open bars were drawn to represent the planned time span for each activity. The figure also shows project status at the end of the sixth month. The shaded bars represent the forecasted span of the activities as of the end of the sixth month. Activity A was completed early. Activity B is forecast to be finished half a month late. Activity C is forecast to end approximately a month and a half early. The percentage of completion for each activity in process is also illu-

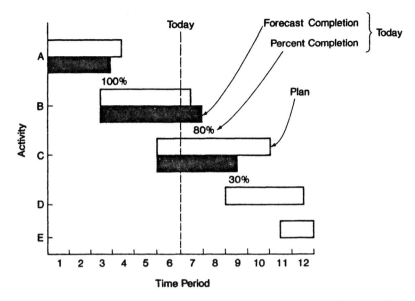

FIGURE 7-3.   Typical bar chart, illustrating a project with five activities at the six-month review.

strated. Activity A has been completed; B is 80 percent complete; and C is 30 percent complete.

Bar charts are simple to construct and easy to understand and change. They show graphically which activities are ahead of or behind schedule. Offsetting these favorable features are some weaknesses, the most serious of which is that bar charts are essentially useless for *managing* a project (as opposed to providing a superficial overview of status). Knowing the status of project activities gives no information at all about overall project status because one activity's dependence on another and the entire project's dependence upon any particular activity are not apparent.

Know the advantages and disadvantages of bar charts.

In addition, the notion of a percentage completion is difficult and is most commonly associated with the use of bar charts for measuring progress (which we discuss further in Part 5). Does the percentage completion refer to the performance dimension, the schedule dimension, or the cost dimension of the job? Unless an activity is linearly measurable (for instance, drilling a hundred holes in a steel plate), it is impossible to judge what percentage of it is complete. (Even in this simple case, the steel plate may have an internal defect, and the last drill hole might be through that defect, causing the plate to crack, at which point what was 99 percent complete now has to be done all over again.) Therefore, percentage completion becomes highly subjective or is frequently taken merely as the percentage of cost expended compared to total projected cost. In neither situation is percentage completion a useful number. Bar charts are much more useful as an indication of what has happened than as a planning tool to aid the new

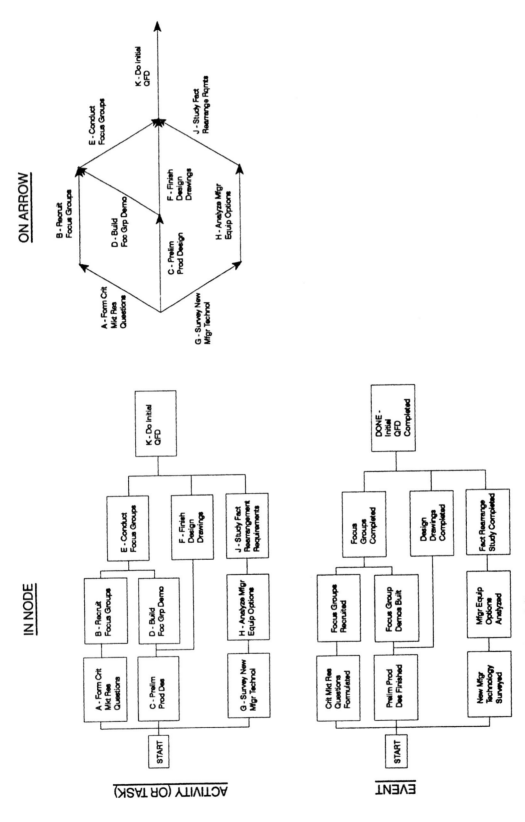

FIGURE 7-4. Principal forms of network diagrams.

product development project manager and the multifunctional project team in making things happen properly in the future.

MILESTONES

A milestone schedule notes a few key events, called milestones, on a calendar bar chart. Milestones have been defined in various ways, but they probably are best defined as events clearly verifiable by other people or requiring approval before proceeding further. If milestones are so defined, projects will not have so many that the conclusion of each activity itself becomes a milestone.

The key to helpful use of milestones is selectivity. If you use only a few key events—perhaps one every three months or so—you will avoid turning milestones into pebbles (sometimes called inchstones) over which people are always stumbling. Some useful milestones might be, for instance, a major design review or a first product test. You may also consider the stage gate (or "go" or "stop") review at the end of a phase to be a milestone.

In common with bar charts, milestone schedules do not clarify activity or task interdependencies. Thus, they must be used with other tools if they are used at all.

A schedule that does not show task or activity interdependencies is useless by itself for planning.

NETWORK DIAGRAMS

There are many types of network diagrams, for which there are a few general symbolic conventions. This topic is treated extensively in many books on project management; we only review a few key points here so you can understand why we favor TBAOA for planning and managing new product development projects. TBAOA can facilitate coordination of the multifunctional project team by clarifying who is to do what by when and who has to work cooperatively.

Types of Network Diagrams

There are many forms of network diagrams, but the program evaluation and review technique (PERT), the precedence diagraming method (PDM), and the arrow diagraming method (ADM) are the most common. A network diagram is any of several displays that link project activities (or tasks) and events with one another to portray interdependencies. A single activity or event may have interdependencies with predecessor, successor, and parallel activities or events. Figure 7-4 compares the three principal types: activity-in-node (AIN), activity-on-arrow (AOA, also called task-on-arrow), and event-in-node (EIN). The following list contains some of these and common abbreviations:

Networks indicate crucial interdependencies.

PERT	Program evaluation and review technique	Event-in-node (EIN)	
PDM	Precedence diagraming method	Activity-in-node (AIN)	
ADM	Arrow diagraming method	Activity-on-arrow (AOA)	
TBAOA	Time-based AOA	AOA with linear time scale	

One of the few useful things a project manager can do when the project is in some difficulty is change the allocation of resources dedicated to an activity. Consequently, if you are not already wedded to and successful with the EIN format, adopt one of the other two. We prefer AOA and, more particularly, the time-based AOA (TBAOA).

TBAOA depicts both interdependency *and* the time schedule. Because so many people are familiar with bar chart schedules, TBAOA is easier for them to understand. TBAOA can be constructed to look very much like a bar chart, but it contains and graphically reveals additional essential information.

Conventions

Activities are always shown as arrows with the start being the tail of the arrow and completion being the barb. Events are shown as circles (or squares, ovals, or any other convenient closed figure). A dummy activity represents a dependency between two activities for which no work is specifically required. Dummies are also used to deal with an ambiguity that arises in some computer-based network diagrams.

Arrows designate activities or tasks. A dummy activity describes a precedence requirement.

A project manager's ability to influence the course of his or her project depends on his or her ability to influence the work on a given task or activity. When the activities are not explicitly shown on a diagram, it is more difficult for the project manager and other members of the multifunctional project team to visualize them and their relationship. One of the few things a project manager can do is change the allocation of resources devoted to a particular activity. Thus, the lack of each activity's explicit visibility in an EIN diagram may be troubling.

A Bar Chart

Figure 7-1 is a bar chart for a new product development project with a planned duration of one year. This shows the intended time spans for each task (or subtask), but not the interdependencies.

An Activity-On-Arrow Network Diagram

Figure 7-5 is an AOA network for the new product development project shown in Figure 7-1. This shows all activities by labels on the arrows and clearly indicates the precedences. The requirement of a dummy activity, a "no-activity" activity, is to indicate that the completion of activities B2 and C2 (as well as activities F and G) must precede the start of activity H.

Figure 7-6 illustrates the next step in using the AOA diagram: redrawing it to a time scale in which the horizontal projection of each arrow is proportional to the amount of time required for its activity. This is a TBAOA diagram. Doing this reveals that one path (B_1, E, F, H) is longer than any other. This is called the critical path. It may also be identified as the path that contains no slack time (amount of time available on a path that is the difference between that required on the critical path and that required on the particular activity path with slack time). Figure 7-6 is drawn with each activity shown starting at the time it was scheduled in Figure 7-5.

The critical path, which indicates the shortest time in which the project can be completed, has no slack.

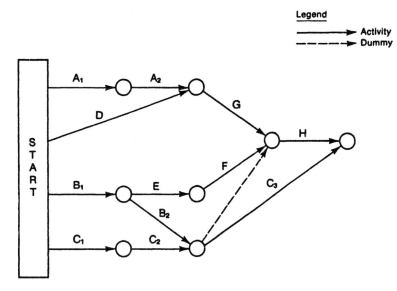

FIGURE 7-5.   Activity-on-arrow diagram with each subtask activity and one precedence condition (or dummy).

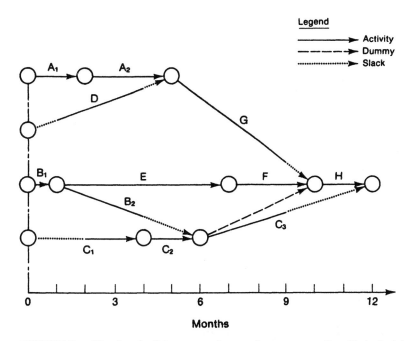

FIGURE 7-6.   Time-based activity-on-arrow diagram, drawn on assumption of task schedules shown in Figure 7-1. (Note that the start node representation is an alternative to that used in Figure 7-5.)

TIME ESTIMATING

Obviously, a schedule for any new product development project requires a knowledge (or estimate) of how long each activity or task will take. Because, by definition, the project has not been done before, there is a necessary inaccuracy in such time estimates. (The only way to guarantee meeting a time estimate is to make it infinitely long, in which case the new product development project won't be authorized.)

Time estimates are usually inaccurate.

After gathering the work breakdown structure and milestones that are necessary to generate a product development project plan, you need to assess their accuracy. How do you judge if the component estimates are ambitious, reasonable, or overly conservative? Although some new product development projects can be estimated based on similarity to past efforts, many are radically different or require development of new technology or processing methods and are thus hard to estimate accurately.

The best you can do is strive to be reasonably accurate in your time estimates. If you estimate a modestly large number of tasks this way, there will be compensating over- and underestimates of time (and cost). The total project estimate will have a smaller percentage error if your multifunctional project team makes estimates for a larger number of tasks, assuming random small over- and underestimate errors for individual tasks.

If some tasks on your new product development project are identical or very similar to tasks previously performed, it is easier to estimate these. (As we said in the previous chapter, one of the goals of breaking the project into small work packages is to obtain understandable tasks, which also implies they can be estimated.) Keep in mind that there are two issues in time estimating. The first is to establish the number of labor hours that participants believe may be *required* for a given task (which will depend on the number, availability, and mix of specific skill levels of the personnel who will do the work). The second is to determine the *elapsed* time for this activity. You must know the former for planning development cost and managing the project, but the latter will determine the overall time to market. However, if a needed resource is not available when required, that shortage can determine the best schedule.

The new product development project schedule depends on the elapsed time to perform all the required tasks.

Pragmatic Time Estimating

We recommend pooled pragmatic judgment for time estimating. Elaborate techniques such as PERT time estimating using three time estimates (optimistic, pessimistic, and most likely) are not justified for new product development projects. In pragmatic time estimating, the task leader, project manager, and one to three others should discuss the task and arrive at a judgment as to what the schedule should be. The task leader is there because of the Golden Rule. The new product development project manager is there to provide balance with other project time estimates. The others are there to bring expertise and experience to bear.

As a practical matter, the project manager, task leader, and one to three others cannot hold discussions on every task on a large project because there simply is not enough time. In such a large project, this is the goal at which to aim, and a practical compromise is for the project manager to have several deputies to represent him or her in these task-estimating meetings.

The goal of such a group estimating meeting is to arrive at a sensible consensus for how long the task will take when it is run in the intended way. If the task leader is a junior person, he or she will not normally be able to complete the task as quickly as a senior person (who may be one of the consulted experts). Sometimes the reverse is true—a junior mechanical engineer may be able to complete a design very quickly using computer-aided design tools, which older product design engineers have never learned to use.

Base the time estimate for a task on who will do the work and how they will do it.

If a task, as distinct from the project of which it is a part, is identical (or very similar) to a previously completed task, then the experience on the prior task is a good estimating guide. But be certain there are no meaningful differences that invalidate the relevancy of previous experience.

A logical sequence for estimating a new task is to (1) determine how many days the previous, similar task required and how many personnel worked on it by consulting existing company new product project records, (2) decide how much more complex the present project is to arrive at a time duration and personnel multiplier, and (3) determine the cost of the new task by multiplying the person days by the appropriate labor rates. This assumes such records exist and underscores the importance of retaining project histories. If no such records are retained, then memory is all that can be used, and memory may differ from one person to another.

What To Do When the Critical Path Is Too Long

In some cases, the logical critical path duration is too long for your company because it shows the new product missing the market's window of opportunity. You normally discover this when you calculate its duration using the natural (or first trial) time estimates for each activity. You can force the forecast completion date to conform to the schedule by reducing all activities in proportion to the time overrun. But this is usually unrealistic.

The schedule must be realistic, and the multifunctional project team must believe it is attainable.

The sensible approach is to replan tasks so there is a credible reason to believe they will be completed in times shorter than the first estimate. In general, start with those that are early in the project, and replan enough of these to achieve a satisfactory completion date. (Keep later project tasks at their initial—hopefully overly generous—time estimate in case you subsequently need to recover from unplanned problems.) In doing this, try to compress tasks that have low compression costs and risks, and be careful not to create another critical path. If this will not save enough time, you will have to replan the project entirely, perhaps running some tasks in parallel. Another alternative is to relax some part of the new product specification so the multifunctional development team is confident it can meet the schedule. If none of these alternatives is satisfactory, realistically you must expect to be late. If that is unacceptable, it may be best to cancel the new product development project and devote the resources to a more attractive—and achievable—new product development project.

COST ESTIMATING

Costs may be stated only in terms of the number of labor hours required, a situation not uncommon in a research group in which a certain number of labor hours have been allocated to a particular project. Cost is more commonly stated in dollars (or yen or marks), however, which entails converting labor hours into dollars. Different hourly rates typically prevail for different seniority levels, and the cost of nonlabor elements (purchases or travel, for instance) is also included. The main elements of any successful cost-estimating system are estimated *labor hours* (perhaps by category) and *nonlabor dollars* for each task in each involved department or group.

Cost is, of course, necessary for planning a project, primarily to justify the benefit-to-cost ratio and secondarily to manage the job. In general, if you intend to compare actual development expense to plan, do not plan costs in detail greater than what you will receive in accounting cost reports. There is no point making cost plans on a daily basis if the company's cost reports are furnished biweekly or monthly. Cost plans, regardless of how they are arrived at, should typically be summarized in monthly periods corresponding to expense reporting. In counting such things as travel cost or computing hours, however, work with hours or days of travel in estimating and sort these into monthly periods. *Plan costs to the level of detail to which they will be reported to you.*

Just as with the schedule dimension plans, there are inaccuracies inherent in cost estimates, and these must be expected and tolerated. But tolerating such inaccuracies does not mean encouraging them. The goal is to be as accurate as possible and to recognize that perfection is impossible.

There is no point in attempting to estimate a budget for an activity until you have established its duration. In addition, the new product development project manager and the multifunctional team should understand the preceding and following activities in order to define better the activity being estimated. Such understanding may clarify that a following activity is farther downstream than it first appears. If so, the activity you are estimating probably is longer, and therefore costs more, than you first thought. *Schedule first; estimate second.*

You do the estimating by breaking the project into tasks and activities, using the WBS and network diagrams. The budget of any large activity is the sum of the smaller tasks that compose it. In general, use as much detail as possible. Every task in the WBS should probably have an individual task estimate prepared by the responsible task manager. *Estimate the cost of each task.*

MICROCOMPUTER PROJECT MANAGEMENT SOFTWARE

Some illustrative microcomputer project software printouts are in Appendix F. Because of the proliferation of microcomputer-based project management software packages, additional forms of network diagrams will probably continut to appear. Many of these project management software packages provide some form of precedence diagram (without a linear time base), but call it a PERT diagram.

Most microcomputer software packages (at the time this book was written) do not produce TBAOA. Many produce box displays with both activity and events in the bases (or nodes). A so-called PERT chart, although portraying interdependencies, lacks a linear time scale and will frequently contain both activities and events within the boxes. Further, you must be cautious because many so-called "time-based PERT" displays are not linear in time. However, several packages do give a true TBAOA. *Be cautious about the features available in project management software packages.*

When and Where To Use Microcomputer Project Management Software

Figure 7-7 provides an overview of management activities and Triple Constraint dimensions showing where microcomputer-based project management software is most likely to be applicable. The big payoff is in planning (or replanning) the schedule and cost dimensions. Its use provides the new product project manager and the multifunctional development team with a "reality test," which forces them to think the project through in sufficient detail to devise some way to complete it satisfactorily. This may not be the optimal plan, but it is a starting point.

Each software package differs in detail. However, they all permit cost estimating when labor and nonlabor resources are assigned to individual tasks. The time schedule for these costs (both expense and capital) can be used as part of the input to the benefit-to-cost financial analysis, as described in Appendix D. The result of this approach is to assure that the new product development's intended project plan and expected financial payback are integrated and—hopefully—realistic.

Because most of this software allows data input in accordance with the project's work breakdown structure, its use is also helpful for planning the performance dimension. If you take the time to update the plan with actual cost and schedule information, this software may also be helpful for monitoring. The critical issue is to stay on schedule; compliance with the cost plan—despite management's normal preoccupation with this aspect—is far less important as long as the company will not run out of money. Finally, as Figure 7-7 suggests, the software will not help you with the definition of what is to be accomplished or its completion.

Figure 7-7 has question marks next to the leading activity under the schedule and cost dimensions. If you are already facile with software and are performing many manual "what-if" trade-offs of schedule and cost, then microcomputer project management software can save you time, which you can use to interact with the people on the new product development project and thus be a more effective leader. Conversely—and this is a substantial trap for the unwary technical specialist—if you sit at a computer terminal performing the "what-if" trade-offs instead of spending sufficient time interacting with the human resources on your multifunctional project team, this software will impede your ability to carry out the leading activity. To reiterate what we said earlier in the book, using project management software is not the same thing as being an effective new

Beware of ineffective use of microcomputer software.

	Performance	Schedule	Cost
Plan	+	+ +	+ +
Lead		?	?
Monitor		+?	+?

FIGURE 7-7.   Best use of microcomputer-based project management software.

product development project manager. Nevertheless, this software will help a good project manager do a better job.

Conversely, insisting that a poor project manager use project management software will not make him or her a good project manager. Similarly, insisting that people working on a troubled new product development project use the software will not fix the project's problems.

This software is certainly adequate today for a wide variety of new product development projects. Because of increasing software and hardware capabilities, newer versions will be able to handle even more complicated and larger projects in the future. Unfortunately, these improved capabilities often are obtained only with the trade-off of greater complexity. Software normally is not required for very small or simple projects.

Cautions

All the microcomputer software packages differ, so there is no uniform set of symbols or standards. Not even the terminology is standard. For example, "time-based PERT" or "time-phased PERT" is typically some form of precedence diagram but frequently lacks a linear time scale. It is unlikely to be the TBAOA we have described.

Keep in mind that microcomputer project software is idiosyncratic. User training is required for every new microcomputer project software package, unlike learning to ride a bicycle or drive a car, where prior experience can be immediately transferred and applied. As project manager, you must assume that your key multifunctional project team personnel do not understand the output diagrams, reports, or formats. Consequently, you must plan to take time to train these people in all this material if you wish them to use it effectively. If your company does not use a single standard package, you may have to teach your managers what the reports and symbols mean. The same is true for outside organizations that are involved, such as subcontractors and strategic partners.

Be prepared to train your key personnel in software use.

Much of project management's challenge arises because people are involved. Microcomputer software is a mechanical device that will not solve the people problems. A project manager cannot sit at a computer terminal, get output, publish it, and expect the multifunctional project team to comply. But if that output is used for discussion with team members and is revised (perhaps several times), there may be a team "buy-in." The key to using microcomputer project management software successfully is to have the multifunctional project team believe that the output represents *their* thinking and for them to understand it. One approach is to use an optical projector to exhibit the computer monitor's display on a large screen for team discussion during a meeting of the key project personnel.

Finally, as with all software, remember the aphorism "garbage in, garbage out" (GIGO). Unless you take the time to enter data that are as accurate and meaningful as possible, your output may be worthless.

Always remember, "garbage in, garbage out."

Bar Chart Formats of Network Diagrams

Although it is often said that network diagrams are difficult to use during project reviews and management briefings because of their apparent complexity, many companies insist they be used for these purposes. There are at least two ways to make management personnel attending such reviews comfortable with AOA presentations. First, activities can be displayed in a bar chart, indicating their planned time, the earliest start and latest finish, and slack, as shown in Figure 7-8. A second approach is to use vertical connections between activities that are dependent on one another to illustrate that dependency (Figure 7-9). However, to use the bar chart representation of an AOA diagram, you must start with the AOA diagram, not with a bar chart. Figure 7-9 is really a TBAOA diagram with all the activity arrows drawn horizontally.

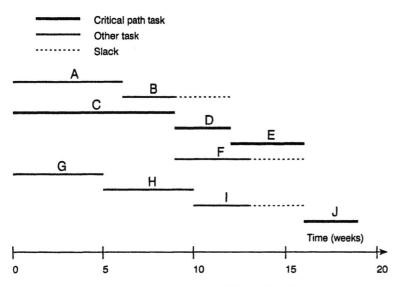

FIGURE 7-8.   Bar chart representation of network diagram from Figure 7-2.

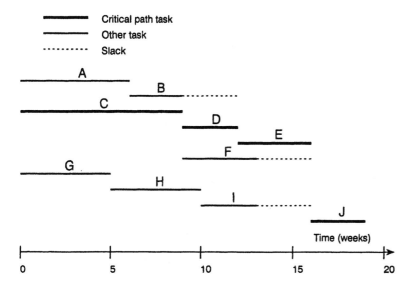

FIGURE 7-9.   Bar chart representation of network diagram from Figure 7-2, with task dependency illustrated.

HELPFUL HINTS

The best way to start is with the work breakdown structure. From the WBS, you can start the network diagram from either the beginning or the end of the new product development project. There are frequently somewhat obvious large subnets you can quickly put down on a piece of scrap paper. As a general rule, it is probably best to start from each end with scrap paper and sort out the connectedness in the diagram where there are activities in progress simultaneously.

Include every element in the WBS in the network diagram.

If you are not using software, you can then transfer the entire diagram to a clean piece of paper. It is probably helpful to do this with a time base and with the presumption that each activity starts at the earliest possible time.

For this initial version, we recommend that you get the people who will be responsible for each activity to estimate how long it will take to carry it out on a normal work basis. When you put these time estimates onto the network diagram, it may become apparent that the entire project will take too long. At this point, you can identify particular activities that may be candidates for time compression, that is, tasks you believe can be done faster.

An alternative to time compression is the parallel scheduling of activities, for instance, software and hardware development. There may be increased risk in such a schedule, but that may be the lesser evil. The final step in creating a TBAOA diagram is to estimate the time of each activity as accurately as possible, using the techniques described in the previous chapter.

Some new product development project (and other) managers resist the use of network diagrams because they consider them complex or because they lack a computer-based project planning network program. This is a serious mistake. It is not the network diagram that is complex; it is the new product development project itself. In fact, if you can't draw a network diagram for your new product development project, that should be a clear danger signal that you and the multifunctional project team do not understand the project. An advantage of using software is that you are forced to confront the logic of your entire project. There is an obvious problem if you cannot load data to get a complete solution, assuming you or a member of your multifunctional project team is able to use a microcomputer.

If you can't make a network diagram, you can't run the project.

We are not opposed to the use of a computer-based planning network program to assist with the mechanics of network usage. In fact, computer programs have great value in determining resource requirements quickly, as we discuss in the next chapter. But a computer-based network diagram program may not be required to manage small new product projects, and the lack of such computer is no excuse for not using a TBAOA.

TYPICAL PROBLEMS

In many ways, the worst schedule dimension planning problem is to avoid the indicated scheduling problems. For instance, a new product development project's completed network diagram may show that required materials will not arrive early enough. This conflict is often avoided or dismissed by saying this can be adjusted later. Maybe it can, but that is hoping for luck to save your project schedule. The solution is to admit the problem exists and revise the schedule to overcome it—now, not when there is no longer time to correct the problem and maintain your schedule.

Sometimes the new product project manager or higher management does not like the forecasted time to market for the project and wants it reduced. Such a reduction is a problem if time is merely cut out of a task without changing the task's work plan to reflect how this reduction can actually be accomplished.

It is difficult to obtain accurate time estimates for things not done before. You must also note overly ambitious or risky approaches so that backups can be identified and planned in advance. As suggested earlier, getting a few people together, including especially those who will be responsible for the activity, and pooling judgments is the best solution to this problem.

The most important problem in planning the cost dimension of the Triple Constraint is that many project groups or project managers have a deplorable tendency to make cost estimates for support group work. This forecloses the possibility of benefiting from support group expertise and violates the Golden Rule. This is easily solved by requiring every department to approve the estimate for the work it will do to support the multifunctional project team.

The correct way to adopt any software is first to specify the problem you are trying to solve with it, second to locate the software that solves the problem, and third to find hardware that runs the selected software. In the case of many people who procure microcomputer project management software, this process is reversed. They have existing hardware, they find some project management software that runs on it, and then they wonder why this does not solve their problem. Your challenge—and your opportunity—is to find and use microcomputer project management software with which *you* can better manage *your* new product projects.

The other major problem is believing that the use of this software is synonymous with managing a new product development project. It is nothing more than a very useful tool to assist with some aspects, and its use must not preclude the new product development project manager from spending time working with the people on the multifunctional team.

HIGHLIGHTS

- Although easy to make and understand, bar charts and milestones alone are inadequate for schedule planning because they do not show how one activity depends on another.

- Network diagrams show activity interdependencies.

- The most common network diagram forms are PERT, PDM, and ADM.

- TBAOA, which can be considered as a bar chart with task linkages, is recommended.

- Time estimating is necessarily inaccurate and can best be done pragmatically.

- Cost estimates for the development effort are necessary but least important if the new product promises an attractive return.

- Microcomputer project management software is readily available to assist a new product development project manager in many important parts of the job.

- The use of this software will require time to train personnel.

- This software will not help the new product development project manager solve many of the interpersonal problems that occur.

CHAPTER 8

The Impact of Limited Resources

This chapter deals with the impact of finite resources on project plans. This topic typically involves two or three dimensions of the Triple Constraint. First we discuss resource allocation and how to resolve resource constraints. Then we present techniques that allow analysis of schedule and budget trade-offs.

RESOURCES

Resources are either people or things. Human resources may include everyone in a particular organizational unit or those with a specific skill (computer programming, senior optical design, or analytic chemistry, for example). Things include any kind of equipment, for instance, lathe availability, computer time, or pilot plant time, as well as floor space to house the equipment and people. Money may also be considered a nonhuman resource.

Allocation

There are three reasons to consider resource allocation in a new product development project. First, forecasted use of some key resource (for instance, circuit designers) may indicate there will be surplus personnel at some future period. This information should warn the appropriate managers either to obtain new business to utilize the surplus talent or to plan to reassign the involved personnel.

Surplus resources waste money and talent.

Another reason for resource allocation is to avoid inherent inconsistencies, for instance, using a particular resource (Jane Draftsperson, for example) on two tasks at the same time. Table 8-1 is a summary of the human resources that might be required for the new product development project illustrated in Figure 7-2 and Table 7-2. No resource overload or conflict is apparent here. When the resource requirements are

TABLE 8-1.   Summary of Resource Requirements for the Project Illustrated in Figure 7-2

Task		Duration (weeks)	Resources		
			Marketing	Engineering	Manufacturing
A	Formulate critical market research questions	6	2		
B	Recruit focus groups	3	1		
C	Preliminary product design	9		2ME+2EE	
D	Build focus group demonstrators	3		2ME+2EE+4Techs	
E	Conduct focus groups	4	2	1ME+1EE	
F	Finish design drawings	6		1ME+1EE	
G	Survey new manufacturing technology	5			1 Mfgr Engr
H	Analyze viable manufacturing equipment options	5			1 Mfgr Engr
I	Study factory rearrangement	3			1 Mfgr Engr
J	Initial QFD	3	2	2ME+2EE	1 Mfgr Engr

displayed on a time scale, it becomes clear that there is too much work for two mechanical engineers (ME) and electrical engineers (EE). They are overcommitted in the tenth through fifteenth weeks, so this new product development schedule can be met only if additional people can be made available. Preparing a network diagram to a time base emphasizes resource allocation and reveals latent conflicts.

A network diagram can show what resources are required and when, which may reveal that more of some resources will be needed than will be available at some time. When you discover this, you must adjust the network diagram to shift the overloaded resource requirement to some other time. If you fail to do this, slippage will occur. Figure 8-1 illustrates resource allocation. In this case, the resource is the personnel headcount. Tasks A and B, each of eight weeks duration, require three and five personnel, respectively. Tasks C, D, and E are not on the critical path, and examination of the earliest and latest times for them shows they can be commenced immediately or as late as the eleventh week. If the company performing this project employs only six people, task D would have to start early enough to be completed before the end of the eighth week, when task B is scheduled to start. If task D starts later than the start of the sixth week, some (or all) of task D will be scheduled in parallel with task B, requiring seven personnel.

A third use of this kind of analysis occurs in a large company. Imagine that tasks C, D, and E are performed by a particular support department, for instance, the design and drafting section. If the design and drafting section was provided with resource allocation information for all projects, as shown in Figure 8-1, they could identify the earliest and latest dates at which the support, in this case, C, D, and E, would have to be applied. Doing the same for all projects would allow the support group to even out its work load and to identify in each case the impact of any slippage.

A project schedule that requires use of already assigned resources is unrealistic.

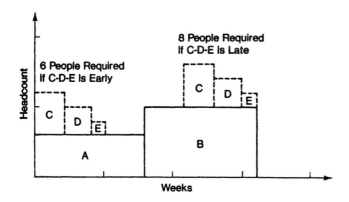

(#) = Required Staff

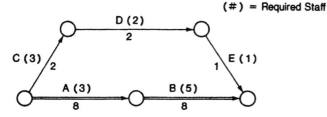

FIGURE 8-1.   Resource allocation.

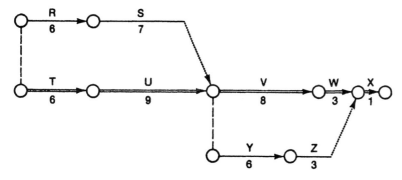

FIGURE 8-2.   Resource allocation with second project. (This project starts at the same time as the first project. The network diagram, including task durations, is shown.)

Multiple Project Resource Constraints

Suppose you have a simultaneous second project (Figure 8-2). Suppose task W requires four inspectors and you have only six. That's no problem on the second project taken alone, but it is when you try to accommodate its need for inspectors with the first project's need, as shown in Table 8-2. Both task W in the second project and task K in the first project are on the critical paths, and they need a total of seven inspectors at the

Lower priority projects usually lose the competition for limited resources.

TABLE 8-2.   Resource Allocation, Using Network of Figure 8-3

Task	Planned Duration	Required Resources					
		Senior Eng'r	Junior Eng'r	Design	Mech	Elec	Inspect
A	3	2	4	4			
B	6		2			3	
C	12*	3	4	4			
D	2*		2		5		
E	11*	2	5		2	2	
F	6		2	7			
G	6	4	1				
H	10	1	3				
J	5		2				
K	1*	2	2				3

* = Task is on critical path.

same time. Something has to be done. Lower priority projects usually get delayed, performance is compromised, or there is a lot of (unplanned) subcontracting. Sometimes it is cost-effective to accelerate a small or low-priority project to get it out of the way and thus avoid a major conflict that would otherwise arise.

Project schedules are usually prepared initially without regard to whether the required resources will actually be available when desired or required. Thus, there can be a serious problem if the impact of resource constraints is overlooked. The first step to avoid this problem is to refine the schedule for your project so that all tasks are consistent with available resources. Then the resource requirements of other projects must be checked and conflicts resolved. These other projects include both those that might start and be in process during your project and those existing (or planned) projects that are supposed to be completed before your project starts but that are delayed to impact your project. If this is not done, the lack of resources will not magically cure itself; it will become an obstacle when there is less (or no) time to devise an alternative schedule.

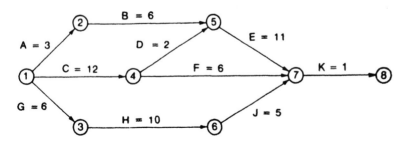

FIGURE 8-3.   Activity-on-arrow diagram.

MICROCOMPUTER SOFTWARE

Most microcomputer project management software will also provide resource histograms. or other information about resource requirements and conflicts. Some of these packages will also automatically level resources or resolve resource overloads. Computers can also do work load prediction and are not prone to making arithmetical mistakes (assuming, of course, they have been programmed correctly and the data have been loaded accurately).

A network diagram is not merely a schedule dimension plan; it also clarifies resource allocation.

TIME VERSUS COST TRADE-OFF

CPM has historically been associated with network diagrams in which there is considered to be a controllable time for each activity. This implies that activities can be accelerated by devoting more resources to them. Thus, there is a time versus cost trade-off for each activity and consequently for a path or the entire project.

Figure 8-4 shows this kind of situation. If you are trying to accelerate a project, you should accelerate the critical path. Of all the activities on the critical path, the most economical to accelerate are those with the lowest cost per amount of time gained.

Consider the following situation that you might face as the project manager for the project illustrated in Figure 8-5.

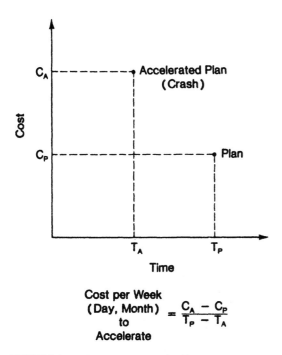

$$\begin{array}{c} \text{Cost per Week} \\ \text{(Day, Month)} \\ \text{to} \\ \text{Accelerate} \end{array} = \frac{C_A - C_P}{T_P - T_A}$$

FIGURE 8-4. Time versus cost trade-off.

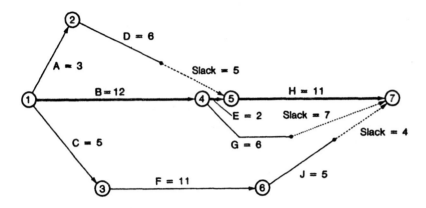

The saving of $200/week can be obtained
for only four weeks, after which path
C-F-J also becomes a critical path.

FIGURE 8-5.   Tie-based activity-on-arrow diagram showing that the saving of $200/week can be obtained for only four weeks, after which path C-F-J also becomes a critical path. (Depending on the certainty you feel for C-F-J and B-E-H, it might be better to shorten only activity B by three weeks, thus maintaining only one critical path.)

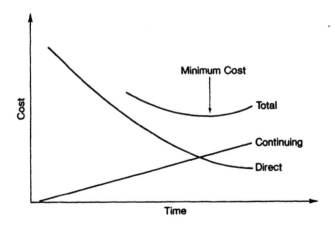

FIGURE 8-6.   Finding the lowest cost.

To carry out this project (assuming all times are in weeks), you must rent a standby electrical generator for the entire duration of the project (however long it is) at a cost of $1,000 per week. Your purchasing department has told you that the subcontractor performing task B has offered to shorten its performance time by as much as five weeks (that is, to seven weeks) but will charge $800 per week for every week less than the original twelve weeks (that is, a premium charge of $800 for eleven weeks delivery, $1,600 for ten weeks delivery, or $4,000 for seven weeks delivery). You can save $200 per week by accepting the subcontractor's offer. However, you have to look at your network diagram before rushing to accept the offer (Figure 8-5).

As you can see, once task B is shortened from twelve to eight weeks (at a cost premium of $3,200, producing a $4,000 saving on the standby generator), there is a second critical path. Thus, you cannot advantageously shorten task B by five weeks, only by four. In fact, you might prefer to shorten task B by only three weeks to avoid having two critical paths.

Figure 8-6 shows another aspect of this. The direct cost curve depicts those costs associated with carrying out the project that are time dependent and for which there is a cost premium associated with shortening the program. In addition, there might very well be continuing costs associated with the program, for instance, the rental of standby power generators or other such facilities. In this kind of situation, there will be a time that leads to the lowest cost for the project.

PROJECT SCHEDULE AND PRODUCT PROFITABILITY

As discussed previously, it is important to evaluate the impact when making a trade-off between project costs and schedule. It is also important to consider more than these internal trade-offs. Frequently, the major impact is *external*—the "cost" of lost product profitability due to a delay. In many cases, the product return on investment (ROI) is very sensitive to changes in product availability (See Appendix D).

Product delay has an impact on the timing and size of revenue projections. It's important to view the market as a fixed "window of opportunity" that appears and then declines as competitive and technical forces move on to new solutions. Under this scenario, it is easier to understand how a delay in product introduction will affect not only the timing but also the magnitude of the revenue stream. The late competitor penetrates the opportunity less and achieves a lower market share. Hence, the double impact (timing and magnitude) to the top line of the income statement can be particularly damaging to the projected financial return for the new product.

TYPICAL PROBLEMS

A major resource problem occurs when project schedules change. This may be caused by a change in your project or another project or by a shift in the start date of a project. Therefore, you must be constantly alert for these changes and their impact on the resources you plan to use.

When first starting a project, the project manager will devise a project plan that details the tasks and milestones along the path to completion. Time estimates can then be placed on each activity to determine the project completion date. These estimates are based on the assumption that the resources—primarily people in the organization—will be available both at the beginning of and throughout the specified tasks. However, as time and the project progress, priorities shift within

(continued)

the organization. When that happens, promised resources are usually moved on to higher priority projects (which often means those that are in bigger trouble). This then results in a change in the project's completion date. Project managers have to anticipate and plan for such resource deficiencies. Otherwise they will often be in the uncomfortable position of trying to justify a schedule slip to the sponsor when it was not due to any sponsor action.

HIGHLIGHTS

- Resources, whether people or things, should be carefully allocated in a project.
- A network diagram can clarify resource allocation.
- Each activity, critical path, and project has a time versus cost trade-off.

CHAPTER 9

Risk and Contingency

In this chapter, we review the sources and residual amounts of risk in projects. Then we discuss contingency, which is a trade-off with risk, and describe ways to insert contingency and reduce uncertainty.

RISK

New product development projects contain inherent risk—perhaps more than most other types of projects. Table 9-1 identifies some of the factors that lead to lower or higher risk. This risk has to be recognized and reduced to the extent possible or practical, and the residual risk must be managed. Table 9-1 illustrates that if the multifunctional project team does not have any contingency (discussed later in this chapter), it (or its company) will bear all the inherent risk. If it inserts infinite or a very large amount of contingency, the company's management is unlikely to approve the new product development effort. Thus, realistic contingency allowances have the effect of allocating the residual contingency between the company's senior management and the project team.

Risk is inherent in new product development projects.

Figures 9-1 through 9-5 illustrate some of the reasons there is schedule risk (and why realistic managers of new product development projects insist on schedule contingency). Figure 9-1 shows the general relationship between the specification's difficulty, resources, and time to market (which is similar to the concept we introduced in Figure 1-5). Optimum resources—which are never available—are exactly the right person or equipment at exactly the right time for exactly the required duration. The best possible schedule, Figure 9-2, assumes optimum resources will be available.

The best realistic schedule, Figure 9-3, happens when the new product development project is supported with realistic resources. These are fewer than optimum, less efficient than optimum, or more than optimum because this imposes communication

TABLE 9-1.   Some Elements of New Product Development Project Risk

	Lower Risk	Higher Risk
Product family	Small modification to existing, well-established product	Requires new technology or is a new market with a new class of customers
Market and competition	Well known, and no or few competitors	Rapidly changing, and many strong competitors
How it is to be made	Well known	Unknown
Duration of project	Short (less chance for the world to change)	Long (global conditions may change) or very short (no time to recover from problem)
Prior experience with similar product	Successful	Unsuccessful
Importance of project to your company's management	Top priority	Unimportant
Background of project and support teams	Pertinent experience and have previously worked together	Lack relevant experience and cooperative work history
Sources of critical components	Multiple, reliable sources or materials	Only one source of uncertain reliability
Project's reputation (if it has been running for some time)	Great	Lousy
Standards	Unregulated market	Highly regulated market
Distribution channel	Existing	New channel required

inefficiencies (and sometimes chaos) on the multifunctional project team. However, the best actual schedule is usually worse than this because, when the multifunctional project team goes to work, its members discover that the specification is more difficult than casual initial examination, as illustrated in Figure 9-4. Regrettably, as illustrated in Figure 9-5, this more difficult specification is sometimes made even more difficult because (1) the technologists insist on including interesting or exciting technology that is not minimally required or (2) the marketing specialists demand some "bells and whistles" that satisfy only a small market segment.

CONTINGENCY

Plans represent the future. Because nobody has a crystal ball, plans must include contingency. Contingency can be thought of as the antidote to risk. The multifunctional project teams that can insert a lot of contingency will bear very little residual risk. In fact, this contingency should be placed on each of the three dimensions of the Triple Constraint. Smart senior managers and executives know that contingency is required and often want to know how much and what kind is included—not to try to eliminate it, but rather to be assured that their own plans (which include your project) can be achieved.

In general, when two people are in a dispute or argument about the schedule or budget for a task or project, the point of contention is not the estimate per se; it is about

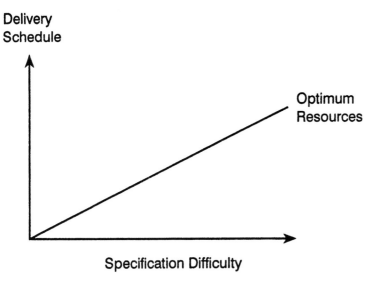

FIGURE 9-1.   Relationship between specification, resources, and schedule.

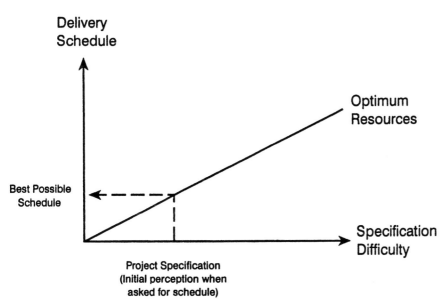

FIGURE 9-2.   The optimum schedule for the initial perception of the new product development
specification.

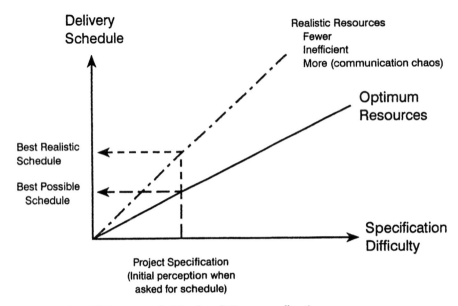

FIGURE 9-3.   The impact on schedule of a realistic resource allocation.

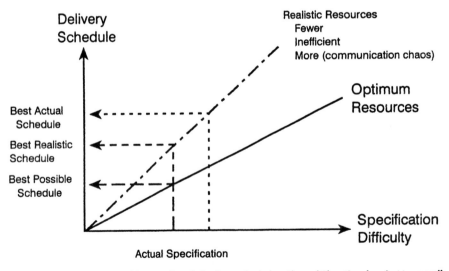

FIGURE 9-4.   The effect of the actual work that is required when the multifunctional project team really understands the undertaking.

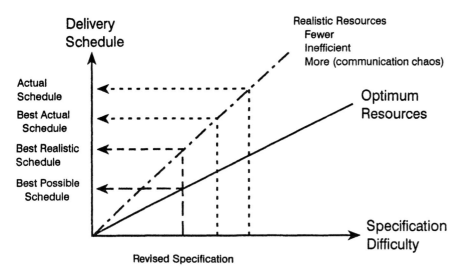

FIGURE 9-5.   The impact on time to market of increasing the specification's difficulty during development.

the risk. Unfortunately, the energy of the argument is usually concentrated on whether the work should take, for example, four or six weeks (or cost $4,000 or $6,000), which cannot be known with certainty in advance. It would be far more productive to discuss what the risks are if the proposed schedule is planned for four weeks rather than for six weeks.

On the performance dimension, it is important that the contingency not take the form of gold plating. Where appropriate, include a small design margin. For instance, if the goal is to have a new car's weight be only twenty-five hundred pounds, it might be appropriate to try to design the car to weigh twenty-three hundred pounds. However, never carry this to an extreme (targeting the car's weight at fifteen hundred pounds, for example).

Build contingency into all project plans.

Contingency is most often associated with the schedule and cost dimensions because projects will inevitably encounter difficulties there. Many things that occur simply require more time and money than planners think. For instance, whenever you must interact with other people, obtaining your boss's approval, perhaps, their schedules constrain you. You will not have instant access, and a delay will occur. If your project involves hiring people, it takes time to train them and make them effective workers on the job. Even if you don't have to hire people, you must still indoctrinate the people who are assigned to work on your project. Similarly, there will be illnesses and vacations pulling people away from the job. These and other tasks, some of which are included in the following list, make it important to build in schedule and cost contingency.

Resource availability
Resource conflicts
Motivation

Something else becomes higher priority
Unanticipated competitive announcement
Sudden price change in the market
Problem with distribution channel
Interface with others
Miscommunications
Get approvals
Get support from other groups
Place major subcontracts and purchase orders
Subcontractor delays
Adjust for labor strikes
Make mistakes
Inefficiency
Train people
Replace sick and vacationing personnel
Cope with personnel resignations
Work at remote locations
Currency rate fluctuations
Overhead and other rate changes
Cope with travel delays
Handle customs duty clearance
Determine whether a new design will work adequately
Advance the state of the art
Accommodate computer down time

Ways to Insert Contingency

There are several ways to insert time and cost contingency. The first is to have everybody who provides an estimate make his or her own time and cost contingency estimates. The problem with this approach is that contingency gets applied on top of contingency, which is then applied on top of other contingencies, and so on, It does not take many multiplications of 110 or 120 percent before the cost of the entire development effort is estimated to be so high that a discounted cash flow (see Appendix D) no longer forecasts an attractive financial return. It does not take many extra hours, days, weeks, or months before the schedule becomes unreasonable. This method is often encountered when multifunctional project teams lack mutual respect between members or are unfamiliar with one another. In this situation, there is an unfortunate tendency to build in protection for their separate functional departments.

Beware of applying contingency upon contingency.

The second method to insert schedule and cost contingency is to put a small amount of contingency, 5 or 10 percent, on each activity in the network. This method is fine, but it misses the point that some activities can be accurately estimated and some others cannot.

A third method is a variation of the second method. The variation, which is far better than the second one, is to explain to everyone providing estimates that they should be as accurate (or optimistic or pessimistic) as possible. Then the entire group can discuss how much schedule and cost contingency should be put on which activities. This can be done by considering the likelihood of things going wrong, the importance of such an outcome, and the maximum and minimum impacts produced by this undesired outcome. There might be some highly uncertain activities that receive a contingency of 50 percent or more. Conversely, a final report might be assigned no or only a small percentage contingency.

Ways to Reduce Uncertainty

Probably the best way to reduce uncertainty is to break the entire project into many small phases, the first of which is a study or definition task. The goal of that first phase is to reduce uncertainty for the remainder of the project. You may also find it helpful to make models, prototypes, breadboards, or other demonstrations. Modern prototyping tools or computer-aided design and engineering software packages may also be helpful.

In addition, using experts or people with relevant experience can be tremendously helpful. Similarly, assuring yourself as project manager that you are the single focal point with total responsibility can reduce communication problems, which are often the cause of major uncertainties.

If the major uncertainties relate to schedule, you can help yourself by starting early. In addition, you may wish to assign additional resources to activities on the critical path. In the case of tasks that have great technical difficulty, you may wish to run a second approach in parallel. Finally, you must of course be certain that everyone paid a great deal of attention to the schedule.

These varied approaches can have an impact on the financial return, so you may wish to model each choice when it is being considered. Padding—the insertion of financial contingency that you do not expect to spend—will merely reduce the apparent return, unless it is really spent, in which case it will reduce the new product development project's actual financial return. When two alternative designs are carried out in parallel, the actual financial return will be reduced, but the likelihood of achieving a high-quality product on time is increased.

External Contingencies

A major project uncertainty that cannot be eliminated is the external market environment. Competitors' product and pricing actions will have a major impact on the new product's acceptance and profitability. Contingency should be factored into the business assumptions to allow some margin for unfavorable market and competitive related developments.

Contingencies represent a dilemma for new product development project managers. On one hand, they have a responsibility to provide adequate contingency so that the chances of achieving the profitability and revenue projections are realistic. At the same time, because the new product is in competition for corporate funds, they cannot afford to allow the product's profit projections to become overly pessimistic.

TYPICAL PROBLEMS

The worst problem concerning contingency is that some managers and custom-ers believe they can save time and money by eliminating it. This belief, or pretense, that contingency is not required restricts the project manager's ability to cope quickly, efficiently, and effectively with the problems that inevitably arise on a project. The lack of contingency reserve (especially time and money) means that the project manager must have a customer or management negotiation every time a problem arises.

HIGHLIGHTS

- All projects should contain contingency, which may be best inserted by adding tasks near the end or by distributing it in each task.
- The amount of contingency is a trade-off with risk.

PART 4

Leading the People Who Work on a New Product Project

CHAPTER 10

How to Organize a New Product Project

In this chapter, we describe the three main forms, namely, functional, project, and matrix, by which organizations arrange their internal reporting relationships and chains of command. Any of these three forms may also be used for a new product development project within an organization. A practical hybrid to create effective multifunctional project teams within a matrix organization is also described.

THREE PRINCIPAL ORGANIZATIONAL FORMS

Projects have a finite life, from initiation to completion. Conversely, a company expects to exist indefinitely. This temporal difference makes it difficult to organize and manage a new product development project within a larger organizational entity.

In addition, new product development and maintenance projects frequently require the part-time use of resources, but permanent organizations try to use resources full-time. Typical project requirements include the following: one hour of computer time each day for a week; one-quarter of Jane Draftperson's time this month and three-quarters of her time next month; use of Joe Technician full-time as soon as the project's circuit designer completes the design. It is important to organize for project work in adequately responsive ways, and it is important for project managers to recognize that frequently this is a compromise that is not fully responsive to project needs.

Although no organizational form is perfect, it is important to support the existence of new product development projects when they are present. This means the organization must plan to accommodate this temporary disturbance and accept some disharmony.

There is a variety of ways that companies or their divisions can be organized and effectively manage new product development projects. The three most common of these organizational forms are functional, project, and matrix.

Organizational forms differ in response to projects.

123

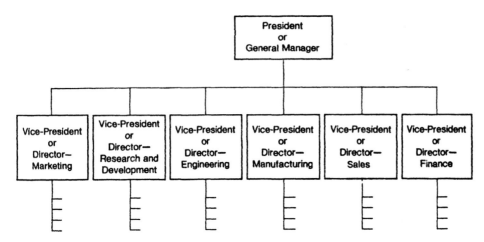

FIGURE 10-1.   Typical organization chart of a functional organization.

Functional

Functional organizations (Figure 10-1) are common in companies dominated by marketing or manufacturing departments (whenever there is a large amount of repetitive work) and exist in other kinds of companies as well. The person asked to manage a project in a company with a functional organization has generally been oriented and loyal to the functional group to which he or she belongs. Specialists are grouped by function, encouraging the sharing of experience and knowledge within the discipline. The functional organization promotes what has been labeled a "silo mentality," in which everything goes in or out of the top and where lateral input and output are nonexistent (or, at least, severely impeded). This favors a continuity and professional expertise in each functional area.

Because such an organization is dedicated to perpetuating the existing functional groups, however, it can be difficult for a new product development project to cross functional lines and obtain required resources. It is not uncommon for hostility to exist between different functions; that is, there are barriers to horizontal information flow, and open channels tend to be vertical, within each function. Absence of a project focal point may trouble a participant interested in understanding the project's status, and functional emphasis and loyalties may impede completion. As a practical matter, the only real focal point is at the top, with the president or general manager, who has far more to do than adjudicate cross-functional conflicts.

From a project management point of view, the functional organization is least desirable.

The functional organization emphasizes functional skills by concentrating these in a small group. Thus, functional experts spend most of their time in proximity to and rubbing shoulders with people of similar skills. Unfortunately, this can isolate them from others with whom they must work if the new product development project is to be successful. Many technical specialists, scientists and engineers, for instance, often lack people skills. This may also be true of other specialists, such as computer programmers.

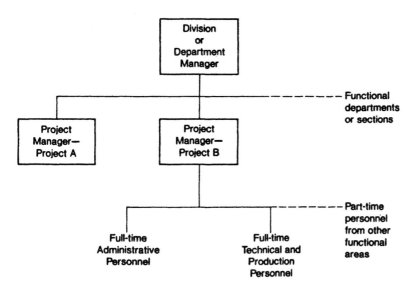

FIGURE 10-2.   Typical organization chart of a project organization.

Because each individual function is best able to communicate with itself, interfunctional communication is often impeded and cooperation hindered.

Project

A project organization (Figure 10-2) emerges from a functional organization when the latter impedes project needs. The solution is to move many of the people working on the project from their functional group to the project manager. Line authority for the project is clearly designated, providing a single focal point for project management. All full-time personnel are formally assigned to the project, thus assuring continuity and expertise.

A major difficulty with this kind of organization is the uncertainty these people feel about where they will go when the project is completed. This terminal anxiety can impede the project's completion. There is also a tendency to retain assigned personnel too long. In addition, it is a rare project that actually has all the required resources assigned to it. Thus, such an organization still requires the project manager to negotiate with the remaining functional organization for much of the required support.

If the organization develops additional projects, managing them in this way leads to a splintering, with many separate project centers existing apart from the functional organization. Duplication of facilities and personnel can result. Managers within the functional organization may feel threatened as people are removed from their functional group. This produces another series of stressors. Project organization often inhibits the development of professional expertise in functional specialties and may not effectively utilize part-time assistance from them.

The project organization form is most useful on large projects of long duration.

Despite the obvious inefficiencies of the project form of organization, this is probably the best form for major new product development projects. If the company has only
one or two new product development projects running concurrently, the inefficiencies
may be relatively insignificant. However, if the company has several simultaneous new
product development projects, some form of a matrix organization normally will be a
better compromise.

The project manager must be a senior person (a "heavyweight" project manager) to
be effective in the project organization. Because he or she has direct responsibility for
the work of everyone involved in the project, the project manager requires direct
supervision of these people. Consequently, project managers in this organization are
often of equal or greater rank than the functional managers.

Matrix

The matrix organization (Figure 10-3) is a hybrid that may emerge in response to the
pressures resulting from inadequacies with a functional or a project organization. It
attempts to achieve the best of both worlds, recognizing the virtues of having functional
groups but also recognizing the need to have a specific focal point and management

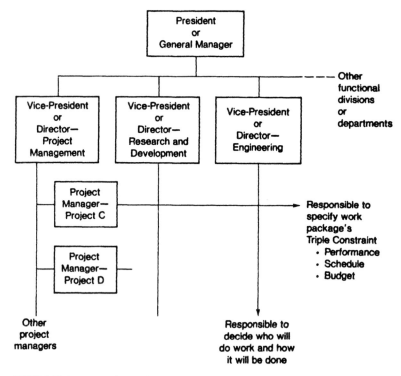

FIGURE 10-3.   Typical organization chart of a matrix organization.

function for each project. Line authority for the project is clearly designated, providing a single focal point. Specialists, including new product development project managers, are grouped by function, encouraging the sharing of experience and knowledge within the discipline. This favors a continuity and professional expertise in each functional area. The matrix organization recognizes that both full-time and part-time assignment of personnel are required and simplifies allocation and shifting of project priorities in response to management needs. Functional departments are responsible for staffing, developing personnel, and assuring the technical quality of the work done by those personnel. The project managers are responsible for defining the work to be done and establishing a reasonable plan (including schedule and budget) for accomplishing it. Project managers and department managers must jointly agree on task, goals, and the specific schedule for each of these.

The main drawback is that a matrix organization requires an extra management function (namely, project management); so it is usually too expensive for a small organization. It is even possible to have a matrix organization within a matrix organization (for example, the matrixed engineering department). In addition, the extra functional unit (that is, project management) can proliferate bureaucratic tendencies, and the balance of power between project management and functional units can exacerbate conflicts.

The matrix is probably the best organizational option if you have many projects.

A matrix organization may be either weak or strong, depending on the project manager's power compared to that of the functional managers. The weak matrix may operate somewhat like a functional organization, and the strong matrix may operate somewhat like a project organization. The project manager's power may derive from financial control, seniority, or simply his or her persuasiveness. Thus, it is possible to have both a weak and strong matrix within a single organization simultaneously.

Project managers in a matrix organization are often concerned about the work of tasks being done by people in functional organizations. Obviously, project managers wish to allow and encourage these people to have as much autonomy as possible to promote synergism, which should enhance the entire organization's efficiency and quality of work. This is particularly true where the functional groups have worked well in the past and have specific knowledge and expertise relevant to the task at hand. Conversely, the project manager is responsible for the ultimate result, must integrate each functional group's contribution, and thus wants to retain as much control and influence over each functional group as possible. Striking a satisfactory balance between these conflicting objectives is very difficult.

EFFECTIVE MULTIFUNCTIONAL PROJECT TEAMS

As is now apparent, there are important reasons to promote both functional and project orientations. Take the use of new factory automation equipment as one example. The key responsibility must lie with the chief manufacturing executive, who is charged with long- and short-term product costs. A single new product's costs depend on both product and process design considerations—best dealt with by a multifunctional new product development team—but also with maximizing the return on investment for any new capital equipment installed in the factory (or factories when there are multiple manufacturing sites).

When two or more new product efforts are under way simultaneously, it is all too easy for each multifunctional project team to make optimum decisions for its own new product development project. However, sometimes a relatively minor compromise on

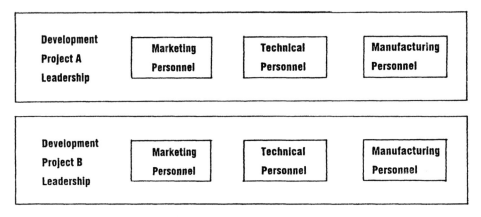

(A)   Personnel assigned to and co-located with the project. About 90 percent of each individual's time is spent on the project.

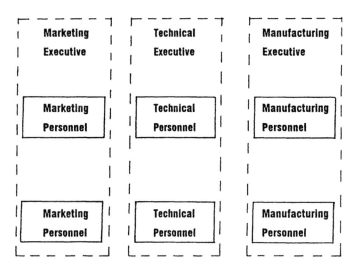

(B)   Personnel coordinate with their functional group executive to gain the benefits of commonality, training, and other skill sharing.

FIGURE 10-4.   One way to achieve strong new product project team leadership and functional group coordination.

the choice of production machinery for each new product development project can produce substantial economies in the factory. Similar situations might occur in other functions: Marketing compromises might reduce common distribution costs, and technical compromises might lead to fewer standard parts with attendant lower procurement costs. Senior management may have to participate in resolving disparate product and functional requirements to achieve a corporate optimum, which is frequently done with the aid of making several financial analyses.

Thus, there is a need for both strong project team leadership and, simultaneously, some coordination with the team members' functional department (or division). Figure 10-4 shows one means by which this might be accomplished. The multifunctional project teams—ideally under a strong ("heavyweight") project leader—are the primary

The project management orientation of a new product development effort has to be balanced against functional organizational concerns and opportunities.

Team Size	Large	Higher intra-team communication "overhead" costs. Investment to promote teamwork required. More experienced project manager required.	Frequently reactive to changes in marketplace. Risk of miscommunication, confusion, innef. efficiency, and wasted money.	Team co-location is often required and justified.
	Small	May require generalists. Less experienced project manager may be adequate.	Normally low cost.	Not a very common situation.
			Frequently dependent on contributions of part-time people. Hard to justify team co-location. Normally not practical to change project manager.	Probably will change project manager as phases shift. Can use full-time, dedicated project team members. Risk of boredom or burnout. Team co-location is possible.
			Short	Long
			New Product Development Project Duration	

FIGURE 10-5.   New product development project duration and team size issues.

focus. However, to assure functional coordination of demands created by other projects, team members devote some small amount of time to issues of their own department. This might be one afternoon each week (that is, 10 percent of the team members' time, as shown in the exhibit); or it might be either more or less, depending on the nature of a company's product mix.

New product development managers play a subtle role. An important measure of their management success is how well they succeed in making the multifunctional team the focal point for the new product development effort. New product development project managers must realize that the unwieldiness of working with a team can have significant long-term benefits. These benefits tend to fall into two distinct categories:

1. Improved decision making that takes into account the viewpoint of multiple functions
2. The long-term morale benefits to the firm that result in team and individual empowerment

There are additional issues to think about when the duration of new product development effort and the size of the multifunctional project team are considered. These are summarized in Figure 10-5. Although not illustrated, the new product development cost is lowest in the lower left quadrant and highest in the upper right quadrant.

TYPICAL PROBLEMS

Each organizational form has its advantages and disadvantages. The only real problem occurs when a new product development project manager believes that a different organizational form will solve all the organizational problems he or she is experiencing. In actual fact, no organizational form is perfect for all situations or for all time.

New product development, whether fast or slow, depends on a multifunctional team effort. However, most companies reward individual behavior, which does not promote teamwork. Customers and users care about the product, not which individuals or departments made it, so functional department managers must evaluate the contributions made to the new product development effort by their subordinates. The company's senior management must establish a reward system to ensure this customer satisfaction.

HIGHLIGHTS

- Three common organizational forms for project management are functional, project, and matrix.

- Any effective new product development project organization must promote multifunctional teamwork.

CHAPTER 11

Organizing the New Product Project Team

The project team consists of those who work on the project and report administratively to the project manager. This is in distinction to the support team (people who work on the project but do not report administratively to the project manager), which we discuss in Chapter 12. In this chapter, we first review sources of project personnel and consider the frequent necessity to compromise by using whoever is available. Then we deal with how much control a project manager can exercise over project personnel and provide some practical tools to help him or her gain effective control. The last section discusses the use of task assignments as a means both to assign the work packages and to obtain commitments from personnel to carry out the work.

DEGREE OF ASSOCIATION WITH THE PROJECT

Table 11-1 shows eight categories of personnel assignment to projects. They result from all possible combinations of three factors: (1) whether personnel report directly to the new product development project manager or are administratively assigned to someone else, (2) whether they work full-time or only part-time each day (or week or year) on the project, and (3) whether they work on the project from its inception to completion or for only some portion of the project.

Project Team

The project team is composed of the people who report administratively to the project manager (the four cells so designated on the left side of Table 11-1). We consider this the project team because the project manager can assign work packages to these people rather than having to negotiate with other managers to obtain commitments for their work.

Team Matrix

The amount of project labor obtained from each category depends on the project's organizational form (that is, functional, project, or matrix) and project size. In a matrix

TABLE 11-1.   The Project Team and the Support Team

Duration of Project Assignment	Nature of Reporting Relationship	Reports to Project Manager		Works on Projects but Reports to Another Manager	
		Works Only on Project	Also Has Other Assignment(s)	Works Only on Project	Also Has Other Assignment(s)
From start to finish		P	P	S	S
Only a portion of project's duration		P	P	S	S

P = Project team
S = Support team

organization, no one may be assigned to work for the project manager; the entire labor pool may be drawn from the support team. In a pure project organization, the vast majority of project labor, perhaps all of it, may be assigned to the project manager. This is especially likely for a large project of long duration. As a practical matter, small projects are not likely to have their own personnel, regardless of organizational form. The majority of those assigned to a project from start to finish is either managerial or administrative because most other skills are required for only some portion of the project.

A key point that emerges from consideration of Table 11-1 is that the new product development project manager must provide eight different kinds of management attention to people working on the project. People who work on the project for only a portion of its duration must be managed to be ready when needed; then orientation to the project must be provided; finally, the project manager must recognize that they may be frustrated or lack a sense of accomplishment at leaving the project prior to completion. People who have other assignments—especially people who are team members on other new product development projects—must be persuaded that the work on the project deserves their attention each day (or week). If everyone who is working on the project is doing so on a part-time basis, other work (sometimes more interesting, but lower priority) intrudes. Because they may have a lower stake in the project, they often require better or more forceful leadership. People on the support team must be managed through other managers, which can lead to distorted work direction. Issues of priorities, performance standards, and loyalty often require the project manager's attention.

There are eight ways to assign personnel to a project; so the project manager must manage differently, as required by each way.

Building an effective multifunctional project team is a formidable but essential undertaking. It goes far beyond obtaining a potential team member's agreement to the assignment. Every multifunctional new product project team will differ in structure, in what is expected of members generally and individually, in the strengths and weaknesses of the individuals and the team as a whole, and in the exact goal of the new product development effort. As another author has observed,

Product design and development is a group activity in which people interact with each other on a daily basis while exploring and assessing options, solving problems, making decisions, taking actions, and seeking assistance. The difference between a cross-functional team that works well and one that doesn't most often lies in the extent to which members have common objectives and are able to create new, shared understandings and meanings. Team members come from different organizational subcultures that may be based on functions, divisions, or geography. Members of each subculture bring their own assumptions, meanings, priorities, ways of thinking, and styles of communicating to the team, and usually assume that their language, styles, and meanings are shared. When these undiscovered differences are not worked through, people leave their meetings and conversations assuming they have been understood. When they discover otherwise, confusion and anger often result, much time may have been lost, and other damage may occur.[1]

Chrysler provides a specific example of some of the difficulties involved:

And the LH team's experience is viewed within the company as a model for changing Chrysler's culture to the core. But, as the LH team discovered, it takes more than throwing a disparate bunch of people into one space to get them to surrender jealously guarded turf and a value system based on individualism for a nebulous concept called teamwork—even when that concept pays off in time and money saved.

The first six months were the roughest. Instead of sitting next to people with the same professional background and similar duties, the 850 LH team members found themselves surrounded by co-workers they had never met doing things they had never heard of. Ill at ease and disoriented, the team made for "a difficult management job," and the work was slow going.[2]

Unless *all* the multifunctional project team members have previously worked together *harmoniously,* team building is likely to be the first chore—and the activity with the highest payoff. Consider the potentially divisive cultural differences between technical and marketing personnel:[3]

Technical personnel	*Marketing personnel*
Introverts	Extroverts
Factual	Conceptual
Detail oriented	Broad brush
Like to know other's qualifications	Accept others as they are
Tolerate failure	Fear failure
Long time frame	Short time frame

Start-up workshops, education, functional transfers, joint events (such as lunches, dinners, or trips), co-location, and the use of organizational development consultants are some of the techniques that the new product development project manager may wish to consider and employ.

A start-up workshop of one or a few days duration—facilitated by an outside expert—is an especially effective method to build teamwork initially.[4] The following are some of the topics worth discussing:

The new product's market and potential competition
Financial analysis (perhaps merely a preliminary one)
New product development project objectives
Intended project management approach

Multifunctional teamwork does not occur by chance. Corporate and project management must encourage it.

A start-up workshop can be very helpful in forming an effective multifunctional team.

Tasks and team members' responsibilities
The schedule
Concepts for follow-on product family members
Key suppliers' roles

It is easier in many instances to achieve highly effective cooperative teamwork in smaller teams because there are many fewer one-to-one interfaces. This implies that a new product development project manager should try to form a team with a few people, each having broad skills, rather than rely on a large team composed of many people with narrow (although expert) skills.

COMPROMISE

It is indeed rare that a new product development project manager can staff the project entirely with personnel who (1) already work for him or her, (2) are enthusiastic about the new product, and (3) represent exactly the right distribution of skills to carry out the project. Usually, the project manager must staff the project from whoever is currently available either full- or part-time. Many of these people will not completely meet the requirements. It is often a case of fitting square pegs in round holes. Unfortunately, in the real world, a project team that has round holes will be staffed entirely with square pegs.

Staffing compromises are usually necessary.

Qualifications

The newly appointed project manager confronted with the urgent need to staff a new product development project team is often victimized by other managers in the company who offer their "cats and dogs." These people may be marginally employed; so company management may pressure the project manager to accept them into his or her group. Consider the problem one project manager called to my attention:

> I inherited an employee who had been with the company for approximately thirty-five years. In all but one or two years of that time he was never properly supervised. He did, however, receive regular raises. He's apparently been allowed to set his own standards of performance, but these are subpar. He has proven to be very devious at beating "the system" with regard to hours and personal activities that are not job related. Now, my attempts at corrective actions and discussions with him produce a very defensive position. He says, "I never knew I wasn't up to par." He exhibits resentful behavior, including mistakes and sulking.

This is a very tricky situation. There is pressure from above to accept the people, and there is another manager offering them as freely available. But, if these people are known to be marginal workers, it is probably better to terminate their employment than to shift them from one project to another, burdening these projects and retaining marginal workers for long periods of time. Nevertheless, it is common for a newly appointed project manager to be offered all kinds of personnel for transfer. On a short duration project, it may be better to accept these workers, unless they are clearly unqualified, than to recruit better qualified assistance.

Motivation

Some new product projects that offer high pay frequently attract workers whose primary motivation is money. The project manager may be besieged by candidates who wish to go to work on his or her project. Their motivation, however, may not be best for the project. So confronted, the manager should seek to staff the project team with a few high quality people and confine the money seekers to support team roles, where they are someone else's problem.

Conversely, a project with high scientific content or one of national importance (such as a cure for AIDS) often attracts highly dedicated, altruistic people. A common correlate of this altruism is a lack of practicality, which the project manager must watch for and temper.

Some projects have an unsavory reputation (fairly or unfairly earned) that makes it very difficult to recruit personnel. They often require portions of the work be performed at an unattractive or remote location. To overcome this drawback, various inducements may be required.

Some project personnel may be poorly motivated, and some may be unrealistic.

Recruiting Qualified Help

Some compromise is clearly required in staffing the project team, but there may be some skill requirements that cannot be compromised.

Most project managers prefer to have people on the project team because it seems to improve project control. People on the project team cannot be given other distracting work assignments unless the project manager approves it, but people on the support team may be given other work that detracts from their ability to honor support commitments.

Thus, in many situations where you want the best qualified help for your project, you must locate a person with the required skill and try to obtain a transfer (perhaps temporary). In some cases, the transfer request will be refused. If your organization has a matrix form, of course, transfer is not possible; at most, you can obtain a firm commitment to assign a needed, uniquely skilled person only as long as the project requires him or her.

CONTROL

Supervision

In talking about the new product project team, we are talking about people who work for the project manager. They may not work full-time on the project, either for its entire duration or full-time within any given workday or workweek. Nevertheless, they are under the project manager's direct supervision, unless there are intermediate levels of supervision. Some of these people may have been transferred from other managers.

Projects go through different phases, which implies that personnel must be changed. For instance, the creative design person, so valuable in the early phases of system design, is not needed when the project is moving toward completion and the team is trying to finish what has been designed rather than figure out additional clever ways to design it.

Thus, some project team people may have to be reassigned during the project. An administrator or junior project manager assigned to work under the project manager may be needed the entire time, but other personnel may need new assignments. They will either go to work on another project full-time after completing their work or work on two or three projects part-time. It is therefore important that the project manager exercises control over the timing of these assignments so as to have people with the right skills available when required and have other assignments for them when they are not required. This is one reason a resource allocation analysis is desirable.

People must join and leave the project as needed.

Proximity

One of the project manager's most powerful tools for improving control of project personnel is to locate everyone in a common area. This aids communication, and, where there is increased communication, there is increased understanding. This also improves the likelihood that everyone on the team understands the Triple Constraint.

Co-locating the multifunctional project team is very desirable.

Problems

Most project managers would rather staff the project with many project team members and fewer support team members. However, this staffing forces the project manager to contend with many personnel problems, such as people quitting (either the company or organizational unit) to work elsewhere, sickness, higher management reassignments, lack of interest in the project, or other conflicting assignments.

Most projects cannot be fully staffed with project team members.

TASK ASSIGNMENTS

We have previously emphasized use of the work breakdown structure and network diagram to divide projects into small pieces of work. Each of these pieces, or tasks, has a corresponding resource and cost estimate. In the ideal world, the person responsible for each task has prepared both the schedule and the budget estimate. This person should also have played a significant role in defining the exact Triple Constraint of his or her small work package. In any case, the project manager must assign tasks to many different people. As these tasks are assigned, some give and take in the exact scope may be accepted, but whatever is finally agreed upon must be committed to paper. That is, there should be minicontracts between the project manager and the people responsible for tasks. The minicontract defines the Triple Constraint of the task.

The project team member who now has his or her task assignment should provide the project manager with a detailed plan of how that task will be performed and periodically review progress against the plan. To the extent that the task performer has played a major role in creating and initiating the task assignment, he or she is likely to be highly motivated to carry it out. Conversely, if the task was assigned without negotiation, the person may have a low sense of involvement and be largely demotivated by the assignment.

All work assignments should be written.

TYPICAL PROBLEMS

The usual problem is what to do with marginal personnel. This is one reason you should have inserted schedule and budget contingency—because sometimes you have personnel who must be used in an area outside their competency, which renders them temporarily marginal. In the case of truly marginal personnel, you can simply refuse to accept them on the project team. You must also expect to work with a group of people with differing and sometimes conflicting personalities. Obviously, these people must work together closely and harmoniously. Depending on the specific situation, you may have to solve individual problems, confront or replace people, or work with the entire group to improve morale and cooperation.

HIGHLIGHTS

- The project team is people who work on the project and report administratively to the project manager.
- Compromise is required in forming project teams.
- People must join and leave the team as required during the project.
- Having team members in close proximity improves the project manager's control.

Notes and References

1. Quoted with permission from Stephen R. Rosenthal, *Effective Product Design and Development.* Homewood, IL: Business One Irwin, 1992, pp. 85–86.
2. A. Harmon, "Teamwork: Chrysler Builds a Concept as Well as a Car," *Los Angeles Times,* 26 April 1992, pp. D1ff. Copyright 1992, Los Angeles Times. Reprinted by permission.
3. Based on a talk by Lester C. Krogh, Product Development and Management Association Conference, University of Minnesota, 16 September 1991.
4. D. Ono and R. D. Archibald, "Project Start-Up Workshops: Gateway to Project Success," paper presented at Project Management Seminar, San Francisco, 17 September 1988.

CHAPTER 12

Organizing the New Product Support Team

The support team is the people who work on the project either full-time or part-time for a part or all of the project but do not report administratively to the new product project manager. This chapter discusses how to obtain their involvement and commitment and how their efforts can and must be coordinated with the multifunctional project team. Then we consider interaction between the project team and support groups and subcontractors.

INVOLVEMENT AND COMMITMENT

As with the project team, the best way to develop a sense of involvement and obtain a commitment from the support team is to have had its members participate in setting the specification and schedule. That is, the more multifunctional project team members who participate in establishing the trade-off between resource requirements (and their realistic availability), the specification's difficulty, and the time-to-market schedule, the greater the buy-in and motivation. Participation also builds a team spirit that continues beyond the project. Failing this, their involvement in planning their own work and committing those plans to writing should also elicit involvement and commitment—remember the Golden Rule: Get the persons who will do the work to plan the work.

Involve support team members in the new product project as early as possible.

New product development project managers, regardless of the specific organizational form of their company, often feel they lack authority. In a project organization, project managers do have authority over the people working for them. Nevertheless, in all situations, as you will see in the next chapter, authority is not especially useful. Persuasion—especially invoking the Golden Rule—is required.

Early Support Group Involvement

Project managers and the project team often ignore support requirements, which other groups must provide, until it is too late. In larger corporations, support groups include such functions as purchasing or procurement, quality assurance, customer service, marketing communications, advertising, order entry, distribution, user training and

138

documentation, international sales, other specialist services, and so on. Unless support personnel understand that their services may be required, they cannot anticipate the extent to which they will be needed. Consequently, the support a project demands may not be available when needed. When support is sought tardily, support groups feel left out, and it may be difficult to obtain their commitment.

This kind of situation may arise because the project team has some degree of parochialism or is not aware what support is readily available. The project team may not understand the potential roles others can play or may assume it knows better than the support groups what kind of effort will be required. This latter situation frequently arises because the project team feels that a support group will "gold plate" the amount of work they propose to do, exceeding project budgets.

As stated earlier, these problems can best be mitigated by involving support groups as early as possible in the project work. Give them an opportunity to participate in planning their task and employing their best thinking and expertise.

Have support team members estimate time and cost of their tasks.

The same applies to the development time and cost estimates. The support group should make time and cost estimates for their task, and the project group should approve them. These estimates may require a negotiated revision to adjust other project tasks to accommodate support group plans if they differ from the project team's first estimate. This is a common occurrence. Support groups sometimes must perform their role at a pace dictated by other, higher priority commitments, thus scheduling your project support differently than you had planned. Sometimes the support group sees a completely different way to undertake its role, often to the project's advantage. Or the support group's experts may convince the project team that their role must be broader than originally conceived. For all these reasons, involve support groups as early as possible.

Written Commitments

Obtain meaningful commitments from support groups within your organization just as you do from the outside subcontractors, namely, a written agreement. (This is also what you should do with project team member commitments; the only difference is what actions you can take to settle disputes that may arise.) There must be a Triple Constraint and signatures by both parties. Such agreements (inside the organization) lack legal standing and enforcement provisions, but if the support group manager must sign a written agreement, he or she will be motivated to a make his or her group live up to its commitment.

Put all support agreements in writing.

Support Team Advantages

As we said in the previous chapter, most project managers seem to prefer to staff their project entirely (or mostly) with project team members. However, a project manager (especially in a matrix organization) might prefer to have a large support team rather than a large project team, for the following reasons:

1. The project manager does not have to worry about the support team after the project ends.

2. In the case of subcontractors, the support agreement is embodied in a legally binding instrument, namely, the subcontract or purchase agreement.
3. The project manager has the whole world in which to find specialists or experts with the required skills.

COORDINATION

Once the support groups have been identified and their work has been planned properly and phased in with that of the project team, there is a continuing need to coordinate project team work. This is best done with network diagrams (Figures 12-1 and 12-2). In both figures, support group work has been segregated from the main part of the network. There are many other ways to do this, for instance, using distinctive line patterns for each support category. Where color copying machines are available, a color code may be used advantageously. Some microcomputer-based project management software allows you to generate schedules for specific groups or individuals. Other software packages allow you to dedicate portions of the schedule chart (for instance, the

A time-based activity-on-arrow diagram aids coordination.

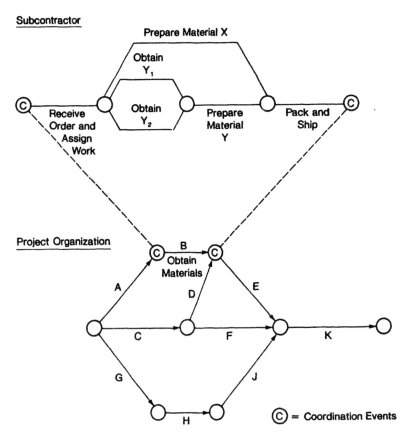

FIGURE 12-1.   Network diagram illustrating use of coordination events.

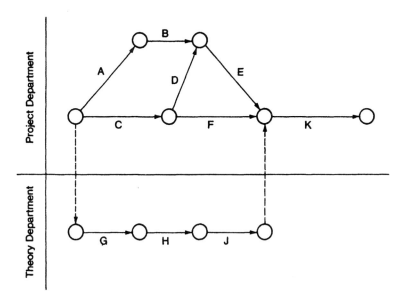

FIGURE 12-2.   Network diagram illustrating use of spatial segregation of a support group's activities.

upper two inches, the middle three inches, or the lower two inches) to the work of a single group or individual.

Change

Consider the network diagram of Figure 12-2. Imagine that you discover you will be late on task D. Should you inform your subcontractor (on task B) of this lateness? It depends. In general, it is probably best to advise subcontractors of your true need date. If you do this, you make it easier for them, and their costs to you will be lower (at least in the long run). But if subcontractors have a history of lateness, it is probably best (1) to have originally allowed time contingency for their work and (2) not to let them know of any delay you have experienced.

Communication and coordination should be primarily in writing. Change should be accomplished by oral communication, over the telephone and/or at meetings involving as many people as required. But the change must then be embodied in the plan revisions.

Revision

Once committed to paper, plans must be disseminated to and understood by all involved personnel. Plans must also be maintained in a current status. If any out-of-date project plans are allowed to remain in circulation, the credibility of all project plans will become suspect. Therefore, everyone who had the original plan must receive revisions. This can be facilitated by keeping an accurate distribution list. Where your company has a local

Write and distribute plan revisions.

area network(LAN), each person (or at least each key person) may have a terminal on his or her desk with the current schedule. Obviously, they must both consult and understand this schedule and any changes that are inserted.

INTERACTION WITH SUPPORT GROUPS

Project team and support group interaction can be difficult. All too often the support group is brought in too late, a situation that reminds us of a story. A commuter comes dashing onto the train platform just as the morning train into the city pulls out. A bystander, observing that the commuter has just missed the train, comments, "Gee, that's too bad. If you'd simply run a little bit faster, you'd have caught the train." The commuter knows, of course, that it is not a matter of having run a little bit faster but rather of having started a little bit sooner.

Project Actions

The tardy commuter's situation classically applies where the purchasing department is involved. Purchased materials arrive later than required; the project is delayed; and project people blame either the subcontractor for delivering late or the purchasing department for failing to place the order early enough. In fact, the blame lies with the people who did not requisition the purchase sufficiently early so the goods would be delivered on time. They need not run faster; they should have started earlier.

Be sure support groups are involved early enough for them to finish when required.

The experienced project manager copes with this problem in two ways. First, he or she makes certain that the network diagram schedule allows enough time for the support groups to perform optimally. Second, the project manager makes certain that all personnel know when task activities must be completed and holds the task managers accountable for meeting the schedule.

Support Team Viewpoint

These issues can be looked at in a different way, namely, from the point of view of the support groups. They are composed of professionals, in the previous example, purchasing professionals who wish to obtain the best quality of required goods at the lowest possible price within the other constraints project personnel impose. They need time to perform their function in a professionally competent way.

The same is true, of course, of any support group—technical writers, computer programmers, designers, draftspersons, or model shop personnel. Everyone wants to do a good job and wants sufficient time in which to do it. But departments have a work load imposed on them by others. They are trying to respond to many projects bringing work to them at random times and in variable amounts. Thus, support groups typically have some backlog they must work through before they can get to new requests. If they did not have this backlog, if they were sitting there idly waiting for the work to arrive, they would not be utilizing a vital organization resource, their own time, in the most effective way.

Support groups labor under many constraints.

One way to improve support (and project) team interaction and assistance is to employ special software. A new category of software specifically aimed at facilitating consensus and decision making among multifunctional team members is evolving. Several vendors, including IBM, currently offer a number of packages specifically aimed at facilitating the team process for activities, including:

- Information sharing
- Brainstorming and idea generation
- Decision support
- Consensus measurement

In addition to facilitating communication among team members, many of the packages are also beginning to include artificial intelligence and multimedia enhancements.

There is more mundane—but nevertheless useful—software support, such as that for part number assignments.[1] This can be used to generate change requests and change orders, for example, and it can also maintain parts status information in a central, accessible file.

SUBCONTRACTORS

Subcontractors are basically no different than you. They have a requirement from a customer, in this case, you or your project. They want to be responsive to you, but they have the same kinds of problems you do: Personnel and resources frequently are not instantly available or perfectly suitable; they need time to plan their work; they have to interpret the Triple Constraint in the correct way, and so on.

Subcontractors can be treated as "them" in a "we" versus "them" relationship or, more effectively, as full-fledged members of the multifunctional project team. Many companies, such as Xerox, have reduced the number of subcontractors dramatically. This may diminish the potential cost advantage of having competitive bids on every procurement of required parts and services. However, it invites and encourages intelligent participation by subcontractors. If they are involved early—for instance, during the QFD and specification-setting stage—they can point out ways to save money or time. Nevertheless, subcontractors ultimately must define their relationship to your project by the contract your company's purchasing department issues to them. Should a change be required, it is certainly all right to tell them about it. But the change becomes effective and meaningful only when it is converted into a contract change.

Another point to consider when working with subcontractors is that your request for a proposal can require that periodic reviews be included in your contract. This is desirable, as it would be if your customer required periodic reviews of your work. You are trying to see how their work is progressing, to understand if changes will be called for as a result of what they are doing or problems they are encountering, in short, to stay abreast of their work.

In many cases, you can do other things to help your subcontractors perform effectively. For instance, you may have one or more lengthy meetings with their project manager to be certain he or she knows what is important to you. You can be willing to compromise noncritical items (with suitable contract changes, if appropriate). You can understand and review their schedule to try to offer constructive suggestions. You can check analyses and witness engineering tests, provided you have made suitable arrangements.

But you must draw a fine line between giving new directions and simply keeping abreast of what they are doing. Remember, the contract dictates the work. The progress reviews or monitoring activities are not a substitute for their management of their work; rather, these are solely to find out if it is being done. You cannot provide daily, weekly, or monthly changes in their direction and expect them to be successful.

Your support agreement is a written contract.

TYPICAL PROBLEMS

Working with the support team probably causes the greatest difficulty, especially for new, inexperienced new product development project managers. The root of this problem is being dependent on nonsubordinates. Two other problems are closely related. First, to negotiate support agreements takes a lot of time, usually at the very busy project inception period. So it is done reluctantly or poorly or even omitted. In the latter situation, the project manager uses his or her own judgment of what the support group will do. Second, even when the support agreements have been intelligently negotiated, later events frequently require that changes be made. Again, this is time-consuming and must be anticipated.

Another problem arises when the support group people are extremely busy and already have a heavy work schedule. In this situation, it is often very difficult to obtain support for the project, and the project manager lacks any direct authority. Obviously, you should try to anticipate these potential problems when the initial plans are made.

HIGHLIGHTS

- Support teams do not work for the project manager in an administrative sense, but their participation and contributions are vital.

- Support groups should be involved in projects as early as possible and allowed to plan their tasks.

- Although they lack legal status, written agreements are an excellent way to obtain commitments from support teams.

- Coordination, a continuing need, is best provided by time-based activity-on-arrow diagrams.

- Every plan revision must be written and distributed to all concerned personnel.

Notes and References

1. D. J. Gardner, "Controlling Documents Can Get Product to Market Faster," *The Business Journal,* 10 August 1992, p. 25.

CHAPTER 13

The Role of the New Product Project Manager

Although the new product development project manager is clearly involved in all phases of the project and is ultimately responsible for satisfying the Triple Constraint, his or her interaction with the project and support teams is a key to the leading (or "people management") phase. This chapter examines the overriding importance of the project manager's ability to influence other team members. His or her leadership ability depends on motivational skills rather than on authority, regardless of how much hierarchical supervisory authority he or she has over project team members.

THE NEW PRODUCT DEVELOPMENT PROJECT MANAGER'S RESPONSIBILITY

Fundamentally, the new product development project manager is responsible for project planning and execution, whereas senior management influences (sometimes profoundly) projects and allocates resources. Distinctions between the roles of senior and project managers emerge in the area of strategy, action, and measurement.[1] We summarize these in this section to clarify the key responsibilities of the new product development project manager in leading the multifunctional project team.

New product development project managers are responsible for project planning and execution.

Strategy

Both senior management and the new product project manager must see to it that the product satisfies the customer and user. They must also both share responsibility to overcome multifunctional project team members' fear of failure while promoting discipline and speed. Senior management must encourage continuous improvement. The effective project manager will be certain that problems are solved prior to management reviews.

Action

Some of the key responsibilities of senior management are to communicate strategy and priorities, designate a proven (or at least prospectively competent) project manager, critically review the product concept and business case or justification, and empower the project team. Once appointed, the project manager must stress the early phases (or stages) of the project because that is where the leverage is greatest. He or she must facilitate broad participation of the multifunctional project team members in negotiation of the key product specifications and design trade-offs. Thereafter, the project manager must buffer the team, resist wasted efforts, and secure required resources in a timely way.

Measurement

Senior management must establish a means to assess the new product development process. They should hold the entire team accountable for the project outcome, while appraising individuals based on their contribution to the team. The project manager must monitor the project status against all its Triple Constraint targets. Throughout the project, he or she should be validating the product's total bundle of attributes to assure it will provide customer and user benefits commensurate with the company's risk in continuing the development effort. Finally, the project manager must promote team efforts that lead to concurrent activity and rapid decision making.

Senior management should hold the whole team responsible for project outcome.

WHAT A NEW PRODUCT PROJECT MANAGER DOES

Influence Rather Than Authority

As the three previous chapters note, many people working on a project do not report directly to the project manager, and he or she does not even have complete control over those who do. In the first place, people are free to change jobs in our society. If given a command they do not like, some workers will simply quit. Or they may transfer to another division of the organization. Second, modern motivational theory indicates that issuing commands is a poor means to encourage people to perform well on a job. McGregor's Theory X/Theory Y is but one manifestation of the thinking that underlies current managerial practice, which usually substitutes persuasion and participation for command. Leadership is exhibited by a person who is followed by others in the conduct of the undertaking.

The project manager lacks control.

Nevertheless, commands are still a way of life, to a greater or lesser extent, depending upon the specific organization and situation. If stated brutally or insensitively, they demotivate and create resentment. If stated politely and reasonably (which is difficult to accomplish), commands may be effective.

Given these limits to hierarchical authority, new product project managers must operate by winning the respect of multifunctional project and support team members. This accomplished, they will find their wishes are carried out voluntarily and frequently with enthusiasm.

One study defines nine influence bases available to project managers.[2]

There are nine ways to have influence.

1. Authority—the legitimate hierarchical right to issue orders
2. Assignment—the project manager's perceived ability to influence a worker's later work assignments

3. Budget—the project manager's perceived ability to authorize others' use of discretionary funds
4. Promotion—the project manager's perceived ability to improve a worker's position
5. Money—the project manager's perceived ability to increase a worker's monetary remuneration
6. Penalty—the project manager's perceived ability to dispense or cause punishment
7. Work challenge—an intrinsic motivational factor capitalizing on a worker's enjoyment of doing a particular task
8. Expertise—special knowledge the project manager possesses and others deem important
9. Friendship—friendly personal relationships between the project manager and others

The first clearly depends on higher management's decision to invest the project manager with power, regardless of power's intrinsic utility. The next five may or may not be truly inherent in the project manager's position; others' perceptions are most important in establishing their utility to the project manager. The seventh is an available tool anyone may use to influence others. The project manager must earn the last two. Projects are more likely to fail when the project manager relies on authority, money, or penalty to influence people; success is correlated with the use of work challenge and expertise to influence people.

The long-range objective of total quality management is the empowerment of the individual. An important step in this process is the empowerment of the multifunctional new product development team. It is vital that both senior management and the new product development project manager recognize that "empowerment" is a two-way street. The most difficult part is to give team members the opportunity to learn from their mistakes because the only path to empowerment is the right to fail. There will be occasions when the project manager must negotiate with team members. A typical approach is to explain the rationale of the effort and to involve the people in planning the detailed work packages. Given this need to influence, an effective project manager must be a superb communicator. He or she must have verbal and written fluency and be persuasive to be effective; the next chapter contains some practical tips on how to improve your communication skills.

Effective Managerial Behavior

As we said in Chapter 1, the project manager must work with people not of his or her own choosing, many of whom have different skills and interests. Furthermore, the project manager is a manager, not a doer. If the project manager is designing a circuit for a new product project, who is planning the work of others? Who is deciding what approach to take to the support group manager so as to obtain the services of the most senior and best qualified person? And who is trying to devise a contingency plan in case the system test does not produce desirable results? The project manager must spend his or her time working with people and planning their work so nothing is overlooked and contingency plans are ready if needed.

On a very small project, the project manager's participation is also required as a worker, not merely as a manager. If not physically, then at least mentally, a project manager in this situation should have two hats, one labeled "project manager" and the other labeled "worker." The project manager must realize which function he or she is performing at any given moment and wear the appropriate hat.

A manager must plan and manage.

Qualifications

Generally speaking, one becomes a new product development project manager because one has been an excellent product manager or product design engineer (or, more rarely, process design engineer) rather than because one has been trained or demonstrated competency as a new product project manager. But a virtuoso technical performance is not a sufficient qualification for managing the efforts of the project and support teams. In fact, one's demonstrated professional skills are frequently problem-solving or technical skills that do not involve an ability to interact with others. But project managers, in common with other managers, need people skills rather than technical skills. Developing people skills can be extremely difficult for many technically trained people who become project managers. Physical systems tend to behave in repeatable and predictable ways; people do not.

The project manager must deal with many intangibles.

Project managers must be people-oriented with strong leadership and superb communication ability. They must be flexible, creative, imaginative, and adaptable to cope with a myriad of unexpected problems. They must be willing and able to take initiative with assertiveness and confidence in the face of substantial uncertainty and, in many cases, when confronted with significant interpersonal conflict. Finally, because projects are temporary, they must have a tolerance for—in a sense—putting themselves out of a job. In short, the role of project manager will be difficult for a narrow specialist (technical or other) or a person who is introverted or inarticulate.

Working with People

Elias Porter has shown that we behave differently when everything is going well and when we face opposition or conflict. People also differ from each other. Some have an altruistic-nurturing orientation; other people have an assertive-directing orientation; and others have an analytic-autonomizing orientation. Although altruistic-nurturing oriented people are usually trusting (a strength), people of another orientation may see them as gullible (a weakness). Similarly, the assertive-directing person's self-confidence (a strength) can be seen as arrogance (a weakness), and the analytic-autonomizing person's caution (a strength) may appear to be suspicion (a weakness). A person with a balance of these orientations, who is flexible, may be seen as inconsistent.

In the case of new product development projects, the variety of required professional skills is exceptionally large. Such projects may involve marketing, research and development scientists, engineers of all sorts, software system analysts and programmers, industrial stylists and designers, and procurement, manufacturing, and finance personnel, and many others. Unfortunately, the training for each of these fields tends to be exclusive, with the result that a specialist in one discipline frequently lacks the knowledge and language fluency to understand and collaborate with people from another discipline. One vital role of the project manager is to assure effective communications between these diverse groups.

To compound this problem, the same thing said to the same person at two different times can produce different reactions. This lack of predictability can be a major pitfall for many prospective project managers. Project managers must deal with both technical and emotional issues. If not already fluent with these human relations skills, they should take a course in behavioral psychology.

The project manager should be chosen because of an interest and skill in human relations.

The project manager sets objectives and establishes plans, organizes, staffs, sets up controls, issues directives, spends time working with widely varied people, and generally sees the project is completed in a satisfactory way, on time, and within budget. The project manager does not do the work of others on the project. A project manager who is an excellent electronic engineer may find it frustrating to watch a junior engineer carry out the circuit design activities on the project. The junior engineer will take longer, make mistakes, and not do as good a job as could a project manager with that technical skill. But if the project manager starts to do the circuit design, it demotivates the junior engineer and lessens the manager's time to function in the most vital role of all, namely, that of project manager.

Managers manage and workers perform the tasks.

Fortunately, total quality management (TQM) and the multifunctional team approach to new product development force project managers into the position of a coach. Post mortems at Hewlett-Packard revealed that the critical factor distinguishing between exceptional and average project managers was the ability to coach and support team members.[3] Thus, the new product development project manager must be able to understand the objective and the decisions, but defer execution to his or her multifunctional project team.

TYPICAL PROBLEMS

Project managers are not normally selected from a pool of trained, qualified people. Rather, new product projects are initiated by a company, and a product manager or technical expert is asked to become project leader.

Such people are often good "doers" and have technical skills and may think they want to be a project manager, but they usually take the job not knowing what is involved. In general, technical specialists rapidly master planning techniques and then the mechanics of project monitoring. They may get along with others (as opposed to being hermitlike) but be unable to cope with the inevitable conflicts that bedevil the project manager. Or they may be poor communicators. What then happens is the organization has a poor project manager and has lost the services of a good technical specialist.

One cure for this problem is to be sure that candidates for new product project management read books such as this prior to being offered jobs as project managers. After that, assuming a continuing interest in the job, the selected candidates should be offered further training.

HIGHLIGHTS

- There are many limits to authority; so project managers should learn how to wield influence.

- Managers must confine themselves to planning and let others perform the tasks.

- Human relations skills are vital to a project manager.

Notes and References

1. Talk by S. R. Rosenthal at the Product Development and Management Association's international conference, Boston, 17 October 1991.
2. H. J. Thamhain and D. L. Wilemon, "Leadership, Conflict, and Program Management Effectiveness," *Sloan Management Review*, vol. 19, no. 1 (Fall 1977), pp. 69–89.
3. J. M. Cusimano, "PMs at Hewlett-Packard Change Ways They Work with Technical Professionals," *PM NetWork*, October 1991, pp. 37–38.

CHAPTER 14

Practical Tips for New Product Project Managers

Because a new product development project manager must be a superb communicator, we first discuss the general problem of communication and then provide several simple suggestions for improving your communication skills. Another pervasive problem for project managers is resolving conflicts, and some techniques to deal with conflict are provided. Finally, we offer some miscellaneous pointers.

COMMUNICATION

Effective communication is one of the more difficult human endeavors. There are so many obstacles it is amazing that any effective communication at all occurs. Words have different meanings, and people often have different perceptions or orientations. The project manager's reputation (be it as a jokester or as a very serious person) will alter the way any message is received. Everyone the project manager communicates with will tend to hear the message he or she wants or expects to hear, which is not necessarily the message the project manager is attempting to deliver. Sometimes people are not listening, are distracted, or have a closed mind. These generic problems are often exacerbated for a new product development project because of the diversity of people working on the effort and the dynamic market forces that often prevail.

Communication must be worked at.

There is an aphorism about how to communicate: First, you tell people what you intend to tell them; then you tell them; and then you tell them you told them. There is much truth in this use of multiple message delivery.

There are several general steps you can take to improve your communication with other people:

Plan what is to be communicated beforehand rather than trying to decide while communicating. As it is sometimes stated, "Put brain in gear before opening mouth."

Use face-to-face meetings in which you can observe the other person's "body language." Allow enough time at an appropriate time of the day.

Decide which sequence and combination of telephone discussion, face-to-face
   meeting, and memo will be most effective.
Be consistent and follow through with actions appropriate to your message.
Use simple language.

In addition, you should consider using feedback, notices, and proximity.

Feedback

Communication is very much like a servomechanism in that it is not effective unless
there is feedback. Communication can be improved by asking the person to whom the
message has been delivered to restate it in his or her own words. This can help overcome
a listener's closed mind. Another effective technique is to back up any verbal communication with a memo. This may also be done the other way around, first sending the
memo and then having a meeting to discuss it. The duality of mode and the recipient's
restatement, rather than being simple redundancy, are most effective here.

Effective communication requires feedback.

Notices

It is impractical to meet constantly with all participants on a very large project. Even on a
smaller project, it may be disruptive to have numerous meetings. It is thus desirable to
issue project notices and reminders of priority actions for any given period. Putting
such notices on distinctively colored paper or preprinting the project name on the top
will set them apart from other mail. 3M's Industrial Specialties Division provides
"FASTRACK" stickers for use by one project per business segment so that these may be
affixed to all project documents[1] In addition, electronic mail and other forms of modern
telecommunications can be used. If you are using project management software with a
local area network, you can deliver many notices through that software.

Proximity

Locating the people on the project near each other also aids communication. Because
the people are close together, they can see each other more often, which makes
communication easier and more frequent. And when people are in frequent contact,
their points of view tend to become more uniform.

   Co-location of the multifunctional project team can be a special challenge for the
new product development project manager. In many companies, to avoid the obvious
and immediate expense of moving people (and perhaps families), people continue to
work at their customary location. However, this choice has a hidden—and often insidious—
cost because scattered work sites cause delays and impose substantial communication
costs. For example, interpersonal communication may be impaired or delayed, and
counterproductive actions may go undetected for some time. The alternative of co-locating
the entire multifunctional project team (or at least the key members) frequently requires
substantial expense. Even if the new product development project manager can successfully justify that outlay, functional managers often resist having their staff members
relocated, especially if the site is remote.

   These issues are more complex when the multifunctional team includes members

There are many issues to the co-location question.

normally located in different countries. Motorola's Keystone pager involved contributions from employees in Boynton Beach, Florida, and Singapore,[2] In this situation, there were cultural and language impediments and a twelve-hour time difference. Although the new product development project was successful, it took longer than intended.

Technology can frequently help multifunctional team communication where geography and different time zones limit the window available for simultaneous communication. The range of software packages from E-mail to sophisticated brainstorming and consensus measuring packages can play a major part in bringing together a widely dispersed team.

On balance, we urge project managers to plan and strive for co-location to the maximum practical extent, especially for critical new product development efforts of long duration.

Follow-Up

It is necessary to have some system of follow-up of the communications, be they face to face or written. Some people simply keep an action log, a chronological listing of all agreements reached with other people for which follow-up action is expected.

Somewhat more effective is a follow-up system keyed to the individual from whom action is expected. A filing card with each key person's name printed on the top may be used to record notations of actions expected of that person. A variant of this is keeping a folder for each key person in which you store records of all discussions or copies of memos for which follow-up action is required or requested. In either event, hold periodic meetings with each key person, and use the filing card or folder to plan the topics to be discussed.

When it is known that project managers (or any manager, for that matter) have such a consistent follow-up system, people who work for or with them will realize that any statements made to them will be taken seriously. Therefore, commitments made to them tend to receive serious and consistent attention.

Follow up communications.

CONFLICT RESOLUTION

Projects are fraught with conflicts. They inevitably arise because projects are temporary entities within more permanent organizations. One root cause is thus competition for resources. Regardless of organizational form, project managers and functional managers tend to have momentary interests that are at odds, so project managers must expect and be able to "stomach" conflict. If you have a low tolerance for conflict, being a project manager can be frustrating.

Two research studies examine the causes of and ways to resolve conflict.[3] Their findings indicate that many things can be done to reduce conflicts, the simplest of which is having good plans, current and realistic schedules, and thorough communications.

The project manager must cope with conflict.

PRACTICAL TIPS

There are a few "tricks of the trade" that project managers should employ to help put the techniques discussed in this chapter to practical use.

First, keep your door open. This encourages people to talk with you, and sometimes that will identify key project issues that you were not yet aware of.

Second, close your door and do not answer the telephone. This is when you do your planning, to gain high leverage on time use. When you are also a worker on your project, which is a common situation on smaller projects, this is when you do your own project work. In addition, a closed door permits private discussions if these are required. Finally, a closed door is a trivia filter, which may force people to stand on their own feet and make minor decisions.

Third, walk the halls. There are always some people who will not enter your office, even if the door is open. Also, what you inevitably see when you go to the sites where work is supposed to be done is that things are not as you expected them to be. At Hewlett-Packard, this is called management by walking around (MBWA).

Fourth, set a good example. Arrive early (or at least punctually) for work; take your job seriously; work hard; and be respectful of others (especially upper management and your prospective customers), even if you disagree with some of their actions.

Fifth, when you are encountering problems, do not try to conceal them. Rather, seek advice from more senior people.

TYPICAL PROBLEMS

The development of your own natural style of management may be a problem when applying the practical tips in this chapter. Our styles of dealing with conflict, for instance, may not be best for you.

Another problem is balancing the two roles of project manager and worker, if the project is such that you must play both roles.

In general, however, the biggest problem is to become a superb, versatile communicator. You must work with varied people—many (or most or even all) not of your own choosing—and that can be done only if you communicate effectively with them. Some may receive and reply to written memos, and perhaps you can write these well. Or (and this will always be the case with some workers) they may not read or understand written memos. So you must talk with and get useful responses from these other workers. Obviously, the reverse may be true also, that is, your writing skill may be relatively weaker than your speaking skill. But some workers receive written information better than spoken; so you may have to improve your writing (or reading) skills.

HIGHLIGHTS

- Effective communication can be aided by feedback, issuing notices, and locating workers near each other.

- If the multifunctional project team cannot be co-located, use all the communication aids you can gainfully employ.

- Conflict between different people (or their group managers) must be expected on a new product development project. Conflict can be reduced by having plans that are current and by good communications.

Notes and References

1. J. J. McKeowen, "New Products from New Technologies," *Journal of Business & Industrial Marketing,* Vol. 5, No. 1 (Winter/Spring 1990), pp. 67–72.
2. S. R. Rosenthal, *Effective Product Design and Development.* Homewood, IL: Business One Irwin, 1992, pp. 222–226.
3. H. J. Thamhain and D. L. Wilemon, "Conflict Management in Project Life Cycles," *Sloan Management Review,* Vol. 16, No. 3 (Spring 1975), pp. 31–50; and H. J. Thamhain and D. L. Wilemon, "Leadership, Conflict and Program Management Effectiveness," *Sloan Management Review,* Vol. 19, No. 1 (Fall 1977), pp. 69–89.

PART 5

Monitoring Progress on a New Product Project

CHAPTER 15

Monitoring Tools

The next managerial activity on a new product develoment project is monitoring progress. First this chapter discusses various monitoring techniques. Then there is a detailed consideration of the use of reports. Finally, we discuss the special case of monitoring several projects simultaneously.

CONTROLLING TO ACHIEVE OBJECTIVES

The word "control" has a pejorative connotation, implying power, domination, or authority. Thus, many project managers (especially new ones) tend to avoid the necessity of installing and using controls on new product development projects. The purpose of such project controls is to measure or monitor progress toward your objectives, evaluate what needs to be done to reach these objectives, and then take corrective actions to achieve the objectives. Thus, you must employ controls (in the measurement sense of the word), or your project will go off course and you might never know it.

Controls are needed to monitor actual progress compared to the project plan.

The ultimate monitoring tool is the product's financial return (IRR, NPV, or ROI, as described in Appendix D). In addition to controlling costs and schedule, the project team should monitor the accuracy of market and competitive information as the project develops. New market data, competitive evaluations, and customer feedback should be factored into sales and profit projections. These estimates are "softer" and not as controllable as development cost and schedule parameters; however, market and competitive assumptions are the major factors that determine the financial return.

CONTROL TECHNIQUES

The first, and in many ways the most important, control is a well-publicized plan for all three dimensions of the Triple Constraint. A work breakdown structure, a network diagram that indicates every element of the WBS, and a cost estimate for each activity indicate how the project should be carried out. Any deviation—and there normally are

What you measure is that to which attention will be given.

159

several—from this three-dimensional plan indicates the need for corrective action. Without such a plan, control is impossible.

There are several restrictive control tools available, such as withholding resources or discretionary authority. When the new product development project manager uses these controls, he or she is assured that people working on the project request the use of these resources or authorities, thus providing visibility. As an example, the project manager could require any expenditure in excess of $1,000 to receive his or her specific approval. Or the project manager could require that any drawing release have his or her signature. These kinds of controls go beyond the project plan in that they make project workers seek out the project manager for approval during the performance of each project activity or task. Anyone's failure to request approval of a planned major purchase tells the project manager that the project has deviated from the plan.

Controls can tell you if the project plan is being followed.

This kind of restrictive control may well be appropriate with an inexperienced team or on a difficult project. But it is normally appropriate only for very small projects. If essentially all decisions on a large new product development project must flow through the manager, the project will get bogged down by his or her lack of time to review a myriad of documents for approval. An effective variation on this restrictive approval control approach is to insist on independent inspection and quality control approvals or on test data as means to verify progress. For instance, you could insist that each subsystem test be approved by people working on other related subsystems.

Another project control method is to place trust entirely in the person carrying out a particular task. This method is fine if that person is able to recognize deviations from plan and realizes they must be reported promptly to the project manager. The person must also be capable of reporting the problem clearly. Because these three preconditions are rarely satisfied, this control tool should not normally be used.

A far better approach is for the project manager to examine the work being done under the direct control of the project team and support teams. This kind of control is based on the Theory Y assumption that people working on project tasks will be trying to do a good job (which can often become a self-fulfilling prophecy), rather than on the Theory X assumption that people will not do a good job. These examinations of activity work are accomplished by reading reports and conducting project reviews.

Control is best exercised by examining the status of tasks.

MILESTONES

A very useful monitoring approach is to focus on new product development milestones, which may be any of several key events in the overall project. This is frequently the choice of very busy senior management, but it is also appropriate for the project manager and the multifunctional project team.

A typical key event is the end of each phase, where the adequacy of the required deliverables is reviewed. At this event, senior management and the multifunctional project team conduct a critical examination of those items required by the corporate new product development procedure to authorize passage through the stage gate to start the next development phase. Other milestones might be at the end of some key tasks, perhaps reviewed only by the new product development project manager. Some other examples might be the agreement on product specifications by marketing, en-

Senior management reviews are common at major milestones and the ends of phases.

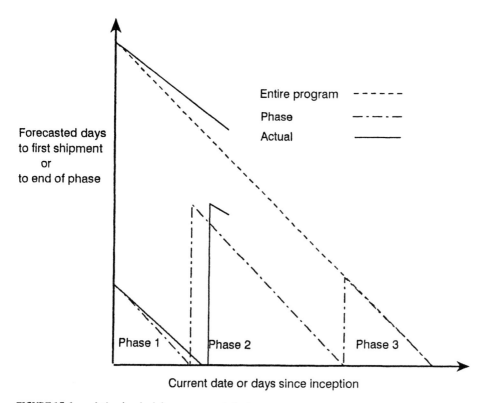

FIGURE 15-1.   A simple schedule measurement chart.

gineering, and manufacturing, electrical test plan completed, all parts qualified, pilot production run completed, successful reliability test, conclusion of beta tests by prospective customers, environmental test report, and so on.

Figure 15-1 is an example of a schedule-tracking chart that might be used at milestone reviews to underscore the importance of time to market. Some companies make the mistake of overemphasizing the development budget by devoting review time to comparing actual project expenses with the planned budget. It's not that this aspect is unimportant, but rather that it has far less importance than the timeliness of the new product's introduction. Unless the actual development expense is vastly over budget or will exceed the cash flow available to the company, the emphasis should be put on the schedule. In a sense, the actual development expense is a measure of the *activity* on the new product development project, not on its *progress*—which can be measured only by the rate at which planned schedule milestones are accomplished.

Do not confuse activity with progress.

Consider the schedule situation illustrated in Figure 15-2. There is a long period of time in the middle of the project when there are no scheduled task completions. Thus, there are no certain checkpoints available, leaving considerable uncertainty as to the actual status. This illustrates another reason to break a project into many small tasks, because there will be less uncertainty as to schedule status.

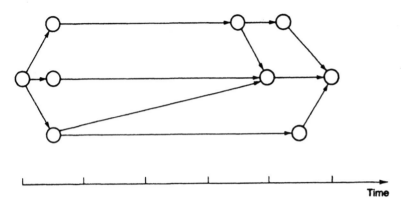

FIGURE 15-2.   Time-based activity-on-arrow diagram.

If the work breakdown structure divides the project into many small tasks, the manager can look at each of them individually and decide whether they are complete. Only those tasks for which all work has been done are considered complete. No other task, regardless of the amount of effort applied so far, is complete. A task that is 80, 90, or 99 percent complete is not complete. This two-state (bivalued) approach to examining tasks simplifies the project manager's job tremendously. He or she can accept as complete only those tasks for which reports (oral or written) guarantee the activity is complete. With many small tasks rather than only a few larger ones, there is much less uncertainty about overall project status.

Dividing a project into many small tasks aids precise determination of its status.

REPORTS

Written reports should always be as brief as possible. Because many people concerned with the project will wish to receive reports, there is a tendency to try to circulate one report to many recipients. This is a mistake. Many who wish to be kept abreast of progress are not interested in small details. This is especially true of busy senior management, for whom the project manager should prepare brief, summary level reports. Such reports might also be circulated to people concerned with some specific aspects of the project but who do not require all the details in any particular report.

Pictures, demonstrations, and models should also be encouraged, especially for a high-technology project or one where much of the work is geographically remote. It is often hard for people not intimately concerned with the project to visualize the status, expected outcome, or even the concept. For them, tangible descriptions, pictures, and such are by far the most appropriate means for providing reports. If your organization has closed-circuit television, this may be an excellent use of it. Some companies transmit computer data by satellite or other relay, permitting a geographically remote work site to construct a model or visualize design information in a three-dimensional display. The ubiquitous facsimile machine also allows people at other locations to see sketches and pictures in close to real time.

It is always important to avoid the pitfalls of excessive reporting. Clearly, this is a gray area requiring judgment. The "convenience" of copying machines can easily lead to too many copies of overly detailed reports being circulated to too many people.

MULTIPLE PROJECTS

As a project manager, you may eventually be responsible for more than one project. You may have many project managers reporting to you or there may be so many task managers reporting to you on a single project that you cannot personally attend all the task reviews and critically examine all the necessary detail. In this case, you must receive some kind of summary information that indicates the status of the several projects (or the many tasks) for which you have responsibility.

When you cannot personally get into details, you are completely at the mercy of those who summarize information for you. When you manage multiple new product development projects, it is probably best to visit the various project reviews personally. Imagine you have three new product development projects reporting to you, each of which is being managed by another project manager, who reviews it monthly. You might sit in on each of these in a rotating fashion; so you attend the review of each project once every three months. If one project you have responsibility for is significantly more important than the others, attend these reviews more frequently (for example, every other month, if not monthly).

Personal reviews are best.

One of the pervasive problems in many companies is that there are too many new product development projects. This overload can produce "gridlock" in which there are too many projects competing for the corporation's necessarily limited resources. So no project has enough resources to maintain its schedule. Resources cannot be effectively hedged over too many projects; senior management must restrict the amount of such activity to those few projects that can be carried out in a timely way. New product development project managers and the multifunctional project teams must share this responsibility by raising a warning flag when they sense there is such an overload.

A company should limit the number of new product development projects so the critical few are not starved of the required resources.

TYPICAL PROBLEMS

Problems arise because of too much or too little monitoring. The former may demotivate personnel, consume too much time, or cost too much money. Inadequate monitoring, based perhaps on the naive optimism that the project will be performed in accordance with plan, can lead to disaster. The following (edited) quotation comes from a memo written months after the start of a new product development project:

A. The accessories were far more complex than anticipated at the time the original plan was prepared. There was not, at that time, any defined level of performance for these accessories; the feasibility of the accessories had not been studied. The estimates

(continued)

for the optical systems and mechanical components of the accessories were inadequate until several months of feasibility study had been performed and the products fully defined. Even after they were defined, all of the accessories have had to go through repeated design iterations. In October, all of the accessory designs were rejected. The Head Adapter and the Lamp Housing required mechanical design rework to reduce their size. In addition, the Lamp Housing required optical redesign to accommodate additional optical features necessary for a smaller, more compact package. The Head Adapter was finally approved in mid-December; however, the base was not approved until mid-January. The Lamp Housing received verbal approval only last week. The Accessory, as designed in this project, was found to be completely unacceptable, requiring total redesign, forcing its removal from this project (all future efforts on the Accessory will be covered by a separate project and are excluded from this summary).

The original plan of last February (13 months ago) contained approximately 600 hours of optical and mechanical design for the Accessories. In our revised plan, including labor to date, the optical and mechanical design is approximately 2400 hours; the difference is 1800 hours plus drafting time.

B. The original plan called for 670 hours for industrial design. Because the designs have been subjected to repeated revisions, the revised plan, including labor hours spent to date, puts industrial design at approximately 1700 hours: an increase of 1030 hours.

C. The layout and design of the Indicator Dial required considerably more time than that provided in the original plan. The Dial design was rejected several times, which added approximately 800 hours.

D. The custom microcircuit for the digital meter version added coordination time with the vendors and with purchasing of approximately 300 hours.

E. It was assumed in the original plan that the proprietary microcircuit would be available as an input to this project. Delays in receiving properly functioning devices made the timing critical. Therefore, coordination with the vendor and evaluation of sample devices of about 400 hours was performed on this project.

F. Due to the complexity of the mechanical design, we added preparation of a tooling plan, preliminary tool design, and coordination to the project. This has added approximately 450 hours. (We will also do the tool design, but that work is covered under a separate project.)

G. The revised plan contains 90 hours for planning and supervision; that time is now charged to the project rather than handled as indirect overhead.

H. The original plan was based on burdened labor rates that were in effect when the plan was prepared. The change in burdened labor rates since then has added approximately 20 percent to the project cost.

I. Personnel being shifted temporarily to other projects has caused approximately 200–300 hours to be added.

J. Approximately 340 hours were added to support engineering time to better match experience on other projects.

K. The original plan allowed for about 800 hours for project administration. The actual and planned hours now represent an increase of approximately 800 hours due to slightly higher monthly hours plus the longer duration of the project.

L. Compared to the original plan, a higher level of technical documentation hours have been expended. Additional drafting time associated with the Accessory mechanical designs and redesigns (due to size rejections) is approximately 600 hours. Other factors were the low estimates for standard parts drawings and reworking circuit layouts due to mechanical and size limitations. These other factors represent approximately 1200 hours.

This project has gotten into severe cost (approximately quadrupled) and schedule difficulties, and it basically got that way by slipping one day at a time. Continuous monitoring could have detected deviations from plan in time to redirect—or terminate—the project.

In addition, there is a monitoring problem because cost (and other) report time lags delay news of project difficulties. In some instances, task leaders or other key people may not be motivated to submit progress reports. This problem is not uncommon where these people are technical experts who are more comfortable with their technology than with writing. Further, there are inaccurate reports, even from conscientious people. In other cases, unclear reports will mask the deviation that has actually occurred.

Also, even if the reports are prompt, accurate, and clear, the deviation may be noticed when you are busy with other urgent activities. Whenever a deviation is noticed, it takes time to react. Finally, you may be distracted by a human tendency to search for a guilty party on whom to blame everything.

HIGHLIGHTS

- Comparison with the project plan provides the basis for monitoring and controlling a new product development project.

- Project managers may exercise control by requiring their approval or trusting task managers, but the best approach is to examine the status of tasks.

- Reports, which may concern any axis of the Triple Constraint and be detailed or general, provide a means to examine the status of tasks.

- Managers in charge of several projects can best exercise control by periodic personal reviews.

CHAPTER 16

New Product Project Reviews

There are two kinds of project reviews: periodic (typically monthly) and topical. Periodic reviews should be convened by the new product development project manager to assure that he or she and the multifunctional project team will actually be meeting to discuss progress. Topical reviews normally will be convened when the multifunctional project team reaches a milestone and requires approval by senior management to initiate a new phase. This chapter deals with both types of reviews and the general necessity to conduct reviews.

THE NECESSITY FOR REVIEWS

Having reviews is very much like having a navigator on an airplane. Reviews and a navigator are unnecessary if everything is proceeding according to plan. The purpose of both is to uncover deviations and correct them. Experienced project managers know the project will not proceed as planned, but they do not know how it will deviate. Only the naive project manager believes the plan is sufficient and no further navigation is necessary to arrive at the project's Triple Constraint point destination.

The project manager's boss and other senior management will frequently want to know about the project status. Although this may not be true for relatively insignificant projects within an organization, it is a very common situation for commercial new product development projects where company money is at stake. The people working on the project will also wish to have reviews of the overall project from time to time. This is their means of learning whether their effort has to be adjusted from plan to conform with some new reality or everything is still proceeding according to the original plan (which never occurs). Reviews with the multifunctional project team in attendance are a means for communication and can enhance their motivation.

Reviews are your off-course alarm.

Senior management must balance their requirements for project reviews. On the one hand, they have a responsibility to stay informed so that resources or redirections necessary to adjust to changes in strategy are implemented in a timely manner. On the other hand, formal reviews consume a lot of time and effort, often two scarce commodities. Also, the multifunctional project team has been empowered to develop the new product.

Excessive reviews by senior management can be counterproductive to this effort. A good general policy is to involve senior management only in those topical reviews at major phase gates when levels of funding or other corporate commitments are made. Major reviews for senior management may also be initiated by the new product development project manager when predetermined levels of change (positive or negative) in the product's anticipated financial return, schedule, or development costs occur.

THE CONDUCT OF REVIEWS

Whether reviews are periodic or topical, they should be planned. Certain questions are almost always appropriate to raise.

Planning

Reviews may be thought of as a very small project. The goal is acquisition of all relevant information. There is a schedule and cost. The schedule may be a simple statement that the review will consume two hours on a particular afternoon. The cost depends on the number of people who participate and the preparation time.

Think of a review as a small project.

There should be a plan for reviews. Everyone involved should understand the Triple Constraint for the review and be prepared to carry out their assignments. This means the new product project manager must know how much time and what level of detail are appropriate for their participation.

Questions

The smart new product project manager learns to ask questions at project reviews. You are not asking questions to embarrass or pillory anyone but rather to find out how the project is deviating from the plan so you can take corrective action. You want to learn about what has and has not happened so far, but you can only change the future. Although questions about the past can help you forecast, they may be intimidating if something has gone wrong. Thus, ask questions nonthreateningly. If you (or others, such as quality assurance) don't ask questions, some people won't volunteer critical information. Such nondirective questions often start with "why." A very good question is, "Why are you doing that?" You can follow this with successive "why" questions. Some other helpful questions to ask are the following:

Ask "why" questions.

What persistent problems do you have, and what is being done to correct them?
Which problems do you anticipate arising in the future?
Do you need any resources (people or things) you do not yet have?
Do you need any information you do not have now?
Are there any personnel problems now or that you anticipate?
Do you know of any things that will give you schedule difficulties in completing your
    task? If so, what are they?
Is there any possibility your task will be completed early?

Is there any possibility that completion of your task will lead to any technological
  breakthroughs for which patents might be appropriate?
Has any work done on your task led to any competitive edge we might use to gain other
  business elsewhere?

The thing to remember about project reviews and these questions is that you are *Plan to conduct*
almost assuredly going to hear some bad news. Most of us do not cope with bad news in a *project review*
very positive way; so the project review can easily become a recrimination and blaming *meetings and*
session. This will not be productive. It will destroy the review and much additional effort *expect problems to*
on the project. Be businesslike and factual in conducting the reviews, and keep asking *surface.*
questions to gather information. If it is appropriate to assess blame, do that in a
different meeting, preferably privately with the person who must be blamed.

PERIODIC REVIEWS

In general, every project should be reviewed once a month by the multifunctional
project team. It may be appropriate to review some projects once every three months;
others may require weekly or even daily reviews. Periodic reviews can catch deviations
from plan before they become major disasters. In the case of the overrun new product
development project discussed in the previous chapter, periodic project reviews could
have caught the deviations at the end of one or two months, when something might have
been done about them, rather than at the end of thirteen months, when the accumulated deviation from the plan was so severe.

Reviews should document changes since the last review or formal report. Documentation of both positive and negative factors that contributed to the change in the
anticipated financial return is an excellent monitoring tool. Identifying current risks
and opportunities for the new product can also be helpful for those outside the
multifunctional project team.

Task Review

A task is in one of two conditions: complete or not complete. For tasks whose performance axis dimension has been completed, examine the actual versus planned cost and
schedule, as illustrated in Figure 16-1. Assuming all tasks were estimated on a uniform
basis, for instance, a 50:50 likelihood of underrun or overrun, and unless there is
something unique about the cost deviations on any completed activity, the accumulated
actual cost versus plan can be used to project the cost at the end of the new product
development project. In Figure 16-1, actual cost for the five complete tasks ($42,000) is
less than planned cost ($44,100), and the ratio of these indicates that the final actual
costs will be approximately 97.5 percent of the plan.

The new product development project schedule normally is much more important
than the development cost.

Actual versus plan ratio may not be meaningful in the case of schedule variations *Watch the schedule*
because many of the completed tasks will not be on the critical path. An activity not on *and recalculate the*
the critical path will often be completed later than plan simply because it was not *critical path.*
necessary to complete it within the planned time. Thus, the project manager can make

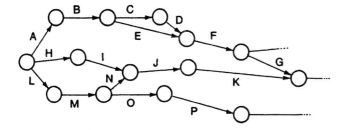

1 - Identify technically complete activities.
2 - Examine actual versus planned cost and schedule:

Completed Activity	Cost			Schedule		
	Actual	Plan	Variance	Actual	Plan	Variance
A	3,000	3,200	200	7	6	<1>
B	4,100	2,900	<1,200>	9	5	<4>
C	6,000	8,400	2,400	6	11	5
D	9,700	12,600	2,900	12	17	5
E	19,200	17,000	<2,200>	24	23	<1>

FIGURE 16-1.   Measuring progress.

predictions about the schedule only by looking at completed tasks on the critical path and then redetermining it. It is also important to watch near-critical paths because these sometimes become critical. Microcomputer project management software can be quite helpful for highlighting the impact of a task's schedule change on the overall new product development schedule.

It is appropriate to ask which incomplete tasks are in progress and which have not yet been started. For those under way, find out whether there have already been any difficulties that would preclude their being completed on time within the cost plan.

In the case of the project illustrated in Figure 16-1, the next concern would be the status of tasks H, L, and F. Do these tasks not yet completed indicate that the project is hopelessly behind schedule? After exploring that, we want to know about the critical path for this particular project. In this case, the concern is whether the completed activities have already caused the project to slip or whether the schedule variations have no significance with regard to the overall project.

Follow-up Actions

During any review, a variety of actions will be identified to cope with the various problems uncovered. The project manager should always record these actions, the person responsible, and the expected completion date. This might be done on a register (Figure 16-2). All concerned people and their supervisors should receive copies, and the

Follow up reviews.

```
┌─────────────────────────────────────────────┐
│        Assigned Actions for Completion        │
│                                               │
│           Project: _____            │
├──────────────┬──────────────────┬─────────────┤
│    Action    │  Responsibility  │     Due     │
├──────────────┼──────────────────┼─────────────┤
│              │                  │             │
│              │                  │             │
│              │                  │             │
│              │                  │             │
│              │                  │             │
│              │                  │             │
│              │                  │             │
│              │                  │             │
│              │                  │             │
└──────────────┴──────────────────┴─────────────┘
```

FIGURE 16-2.   Typical action follow-up form.

status of these action assignments should be reviewed no later than the next project review.

TOPICAL REVIEWS

There may be many kinds of topical reviews for new product development projects. Topical reviews are not held at fixed time intervals but, rather, when certain events have occurred or specified activities have been completed. The specific reviews and their titles will depend on the company's new product development procedure and should be clearly shown in the explanatory roadmap overview.

Topical reviews are event based.

In general, a topical review will be held at the end of each major phase, and the review is frequently a precondition to starting the subsequent phase. As such, these are often called stage gate (or tollgate) reviews, implying that they are a gate through which you must pass to enter the next phase. The topical reviews might be as follows:

Concept
Development

Design
Manufacturing
Launch
Postcompletion audit (or post mortem)

In the case of Keithley Instruments' new product introduction process (Figure 3-4), the topical reviews would be "concept proposal," "study proposal," and "authorization." The exact titles for and number of topical reviews will vary from company to company and will differ somewhat for software and hardware new product development projects.

There may also be other topical reviews that are not stage gate reviews. These might include a preliminary design review, a critical design review, or a preshipment review. All topical reviews should clearly be designated as task activities on the network diagram, and their completions are major project milestones, as discussed in Chapter 7.

Topical reviews conducted for senior management are often onerous, but they may stimulate participation and involvement on the part of all members of the multifunctional project team. For this to happen, the project manager must solicit ideas from all personnel as to what should be discussed during the review. Obviously, some of the agenda will be set by senior management or the new product development procedure's requirements. Rough ideas for the review should be delineated in a smaller group of key staff.

At this point, the project manager should delegate portions of the topical review to members of the multifunctional project team. (Be sure there are not so many people making presentations at the formal review that it becomes a circus.) Next, conduct a trial run with a fairly large group of the team invited to criticize. It may then be helpful to conduct a second dry run with a peer management group that represents the same range of skill backgrounds as the senior management audience. The people attending this dry run review will provide additional insights as to how materials can be better presented or changed. Now you are ready to conduct the formal topical review, after which it is desirable to conduct the entire review again for the benefit of all members of the multifunctional project team.

Topical or stage gate reviews are a crucial element of an effective new product development process.

Include all required topical reviews in the new product development project plan.

Try to involve the entire multifunctional project team in preparation for a topical review.

TYPICAL PROBLEMS

Reviews are plagued by three common problems. First, there is always a concern as to whether the information being presented or discussed is accurate. Beyond this, there is often speculation about the exact status of some task, test, component, or report. Clearly, good planning for reviews can reduce if not eliminate this problem.

The second problem is the poorly conducted review. Aimless discussion or recriminations are common. Running a review like any other well-planned meeting greatly reduces the possibility of getting off the track.

(continued)

The third problem is that some project personnel fail to meet schedule deadlines and do not consider this to be a major problem. For example, technologists may believe it is more important to arrive at a perfect or elegant solution (however long it takes) rather than an adequate solution (on schedule). "Better" is the enemy of "good enough."

HIGHLIGHTS

- Reviews uncover the inevitable deviation from plan and allow a consensus as to the needed corrective action.

- Reviews, like projects, must be planned.

- Ask nondirective questions, and expect problems to surface at project review meetings.

- Periodic reviews should be conducted as appropriate for the project, but once a month is a good rule of thumb.

- The kind of topical review used depends on the project requirements.

- Questionable accuracy and poor procedures are common problems with reviews.

CHAPTER 17

Handling Project Changes

Change is a reality of new product project management. This chapter first reviews why project plans are altered, suggests how to solve the problems created by the change, and then discusses techniques for making plan changes.

REASONS FOR CHANGES

Deviations from the project plan occur because Murphy was an optimist. The new product development project manager and his or her multifunctional project team can reduce the likelihood of deviations by carefully creating a sensible project plan. Optimism or competitive pressures, however, often induce these conscientious people to be too ambitious. Thus, the performance goal, development schedule, or development budget is unattainable. Although the latter dimension of the Triple Constraint is the least important, it is not insignificant. At some point, a development budget overrun can diminish a project's prospective financial return to an unacceptable level. As we have said elsewhere, the crucial dimensions are the performance level (which includes the factory cost target) and development schedule. Changes to these aspects frequently signal that the rationale for the entire effort has become questionable; so the project manager and the team must be especially alert and self-critical in watching progress (or the lack of it) here.

Change is a constant on projects.

The multifunctional project team must also be alert to changes in the external environment. They must constantly seek information that might invalidate their business and market assumptions. Changes in unit sales or price must be addressed realistically and updated regularly. Often it is easy to maintain unrealistic sales or pricing estimates and avoid the impact on the product's prospective financial return. Customer input and constant "reality checks" are necessary to maintain an attainable business case.

The situation is somewhat worse than average for large projects and better than average for small projects. This seems to make sense; intuition tells us that the more ambitious undertaking is less likely to be estimated accurately. This is another reason to break a large project down into many small tasks. It will be easier to estimate a small task accurately.

Environmental, health, and safety regulations that change during the course of a project may cause changes in scope. Inflation may exceed plan, causing a cost problem, particularly on projects originally planned to take several years. It is almost axiomatic that there will be changes in resource availability, either people or facilities. These do not constitute changes in project scope, but they do constitute changes from plan that will have an impact, usually unfavorable, on the new product development project's schedule and cost.

THE GENERAL APPROACH TO SOLVING THE INEVITABLE PROBLEMS

In general, the options available are either deductive or inductive logic. In the former, the solution is derived by reasoning from known scientific principles, using analytical techniques, and the conclusions reached are necessary and certain if the premises are correct. In practice, the project manager is rarely confronted with problems for which this approach is appropriate and must rely on inductive techniques, for which the scientific method is the typical prototype. Inductive methods reach conclusions that are probable. This straightforward approach entails seven steps, described in the following sections.

Good solutions require a seven-step approach.

State the Real Problem

The key to problem solving is understanding the real problem rather than the apparent symptoms. Smoke may be emerging from the hardware you built or the computer may refuse to obey a subroutine command, but the actual problem may be an overheated component or an improper line of code in the computer program. You will have to decide how and perhaps why these particular problems occurred. Take a wide view of the apparent problem to separate the true cause, which is often obscure, from the readily visible symptoms. For example, there is a tendency to blame price for many market-related problems. Don't provide a knee jerk technical solution to a nontechnical problem.

Gather the Relevant Facts

A fact-gathering phase is usually necessary to clarify the problems. People trained in engineering and science tend to want to engage in this step ad infinitum. This has been called "analysis paralysis." Although it may take a good deal of time to locate information sources, there is also a law of diminishing returns. Because you will never have a 100 percent certainty of obtaining all the information, you must learn to exercise judgment as to when to truncate a search for additional information. At that point, you begin to converge on a solution using the information already gathered.

Propose a Solution

Once a plausible or possible solution has been identified, the winners and losers rapidly separate in their approach to problem solving. The losers inevitably adopt the first solution that comes to mind, possibly leaping out of the frying pan and into the fire. Admittedly, the pressures to come up with a solution quickly are great. No one likes to walk into his or her boss's or customer's office and say, "We have a problem." Such a crisis generates psychological pressure in the project manager to come up with a solution quickly so he or she can say, "We have a problem, but don't worry about it because we have a pretty good solution in mind." But it is best to take a different approach.

Develop Several Alternative Solutions

The winning approach to problem solving is to develop several alternative solutions. To quote Alain (Emile Auguste Chartier), "There is nothing more dangerous than an idea, when it is the only one we have." Thus, when the problem has arisen and must be reported, the successful project manager will say, "We have a problem. We may have a possible solution, but I am going to take three or four days to consider other alternatives. Then I will report to you on the options and our recommended course of corrective action." Although such an approach to reporting the bad news may make you initially uncomfortable, it is invariably associated with reaching better solutions.

Developing alternatives is the key to problem solving.

Adopt the Best Alternative

After deciding what is the best alternative, you must adopt a course of action.

Tell Everyone

As an effective project manager, you have earlier made certain that everybody involved in your project knew the original project plan. Now, because you have changed one or more dimensions of the Triple Constraint, you must tell everyone what the new plan is. If you fail to do this, there will be some people working in accordance with obsolete direction, producing something useless and out of date.

Audit the Outcome

As you implement the best alternative solution, watch how it is working out. Auditing will improve your ability to solve problems by showing you how your solutions actually work out. And as you learn more about the problem you are solving and the approach you have adopted, a better alternative may become clear, which may necessitate a further change in the plan.

ADOPTING CHANGES

At this point, it should be clear that projects normally require changes of plan, although the specific reason or reasons cannot be forecast. The original plan should have included contingency for each dimension of the Triple Constraint, as discussed in Chapter 9.

Changes, just like originally intended work, must be defined, planned, managed, and monitored before they can be completed. Thus, some, if not all, of the originally issued task authorizations (see Chapter 6) must be changed when a change has occurred and a decision has been made as to how to alter the plan to carry out the remainder of the project. This may seem to be a lot of work, but it is far less onerous to take the time to make sure each agreement with people working on the project has been changed than to discover later that some people have been working according to their prior understanding of the project plan.

There is a natural reluctance to make a formal plan change. Such a change not only requires work to issue plan revisions, but it forces us to admit we were wrong (in the original plan), and often brings this to the attention of higher management. Conversely, such information, if given to higher management, may result in your project receiving more help, better access to resources, or higher priority.

Never hesitate to publicize plan changes if these are required.

MICROCOMPUTER SOFTWARE

If the changes that have occurred on your project do not alter the logic of the plan, merely the duration or cost of each activity, it is easy to use microcomputer software to determine and visualize the impact. Conversely, when the changes require a substantial revision of the plan's logic, it may be more trouble to revise the plan in the microcomputer than to start afresh and merely plan the remaining work. It can be very time-consuming to attempt to disconnect task linkages that have previously been entered and replace these with new tasks with a different logical relationship between them.

TYPICAL PROBLEMS

If you change project (or task) managers during a project, you will almost always be confronted with changes, typically unfavorable. The new person may be unfamiliar with the ultimate user environment, although that is not always the case. Similarly, the new manager may not be familiar with team members' capabilities. The overriding issue, however, is that the new project or task manager did not plan the work for which he or she is now responsible—this is a violation of the Golden Rule. If you must change the project or task manager during the course of the activity, allow that person some time to review the prior plan and try to determine how to carry out the remainder of the work.

There is always a reluctance to tell the customer and your boss that a revolting development (such as the discovery of unexpected noise in an amplifier) has occurred and many reasons to justify delay. But you should deliver the bad news carefully, thoughtfully, and promptly, before someone else does it.

A second problem is that task authorizations are often verbal rather than written. Because they promote misinterpretation, verbal authorizations should be avoided. But they are employed in the real world of project management. When you must use them, be sure you are clear, ask for feedback, and then try for written confirmation.

The third problem with changes is their impact on resource allocation. There is nothing to do but face up to the reality that resources must be rescheduled, as inconvenient as this may be.

In most project management problem-solving situations, it is not possible to find *the* answer, only a most acceptable (or least objectionable) answer. This may be caused by the inherent uncertainty or lack of quantitative data. It is thus a matter of judgment about when to choose among the identified solutions and when to keep looking for more, better solutions. Honest people will differ (as they will in their perception of the problem and their evaluation of solution alternatives), and this must be both expected and tolerated.

HIGHLIGHTS

- Changes will occur on every project.

- Authorization documents can be used to communicate planning and change control.

- Three problems changes can cause are managers may be reluctant to inform senior management of them, verbal authorizations often cause misunderstanding, and resources must be reallocated.

- The seven steps in problem solving are identify the problem, collect the data, devise a solution, search for alternative solutions, adopt the best solution, implement the solution, and audit the outcome.

PART 6

Completing a New Product Project

CHAPTER 18

How To Complete a New Product Project

The fifth and last managerial activity is project completion. In this chapter, we discuss the consequences of project completion, and show that all personnel do not necessarily have the same stake in ending the project. Projects end with their completion, but there frequently are postcompletion activities that are necessary to the project and that may be viewed as part of it.

ENDING THE DEVELOPMENT

Termination

There is a variety of ways to stop projects. Resources can be withdrawn, for instance, by reassigning personnel or required facilities. Higher priority projects may gain at the expense of a low-priority project, which may be allowed to wither on the vine. These approaches are not as desirable as an orderly and carefully planned termination. Project success, that is, satisfying the Triple Constraint, can be obtained only by this latter approach. Sometimes senior managers become greedy and keep initiating new product development projects without examining resource constraints, so some projects lack the people or facilities to be completed successfully. In some cases, the initiative to initiate a "bootleg" new product development project is taken by an enthusiastic champion. Regrettably, in many instances, these now lower priority projects are not stopped formally. Rather, inadequate resources are still applied, although there is virtually no likelihood of achieving a high return on investment. These same resources might have helped speed another new product to market.

The best termination is orderly and carefully planned.

The key element everyone must keep in mind is to apply *adequate* resources to only as many high-payoff new product development projects as can be successfully completed. If a new, better opportunity is suddenly recognized, stop (temporarily or permanently) a lower priority effort to assure that the newly identified new product development project receives adequate resources to be successfully completed. From time to time, it might even be better to delay—only briefly, of course—the start of a new high-priority new product development because that would permit the orderly successful completion of a previous project that was nearly finished.

Acceptance

The goal of the new product development project manager and his or her multifunctional project team is to obtain widespread—and hopefully enthusiastic—market acceptance of the new product. The purpose of staged new product development is to refine the design, manufacturing, support, and marketing requirements for the product, not to redefine or improve the basic specifications. The basic product must reach the market in a timely manner. Trying to get it "too right" the first time can be a well-meaning but destructive tactic. Remember, the concepts of continuous improvement and the product family are the only reasonable way to address the needs of an uncertain and constantly changing marketplace.

Market acceptance of the new product is the goal.

To repeat what we said earlier, the "product" might be only a tangible entity, or a service, or a combination of the two. It is user satisfaction with the total offering that distinguishes successful acceptance from a marginal result or failure. This means the customer agrees that the performance dimension specification of the Triple Constraint has been met. Unless the acceptance criteria have been clearly defined, there may be discord at the end of the project. When agreement is lacking, some company personnel— frequently sales or marketing—will argue for more performance at a lower product cost. If they fail to win the argument at this late stage, which is the normal outcome because the product features and production processes were established previously, they may resort to destructive blaming and criticism.

Therefore, the acceptance phase must start with the initial written definition of the work to be undertaken. This is the overriding benefit of having a multifunctional project team use QFD during the early specification-setting stage. The goal is to get the inputs from and agreement of *all* the key players before there is either a psychological or substantial tangible investment in a defective or deficient product design specification or approach.

In some projects, it may be impossible at the beginning to agree upon final acceptance criteria. This is typical in commercial product development projects, where marketing is reluctant to settle on all the performance specifications. When this is the case, the product specification should call for an initial effort of an adequate duration to clarify the entire product design and acceptance criteria. At the end of this first phase, a review is conducted, and a firm product specification is agreed to for the final phase of work, including acceptance criteria for the end of the project. This is, of course, the benefit of using a phased or staged new product development procedure, as we pointed out previously. Although we do not recommend it, there may even be several refinements in the product definition at various stages—if that is the way your company has chosen to establish its new product development roadmap.

Completion requires objective and measurable criteria be attained, which ideally satisfy the customer's needs.

Project completion clearly depends upon the precise wording of the acceptance criteria. There should be no room for doubt or ambiguity, although in practice, this is extremely difficult to accomplish at project inception.

Clear acceptance criteria are required.

Transition from New Product Development to Current Product Support

Once routine product shipments commence, the *new* product becomes a *current* product. The sales, distribution, support, and manufacturing functions now become primary.

In the case of a product developed for an original equipment manufacturer, delivery may or may not be completion. Project completion often requires that the product function after delivery at a location the customer designates. Even if not explicitly the case, there may be an implied warranty that requires it. Thus, responsibility for delivered goods after they leave the company is always an issue to be considered at project inception of an OEM agreement.

Documentation Reports

It is not at all uncommon for a new product development project to require the delivery of documentation as well as some tangible output. Such documentation might include a spare parts list, instruction manuals, and a list of service centers, for example.

It is frequently difficult to complete documentation at the end of the project for two rasons: First, many technical specialists are poor writers or are reluctant to write. Second, in many instances, the people who have essential knowledge have long since been assigned to some other activity and are no longer working on the project. The second problem can be eliminated and the first problem diminished by what we call "creeping documentation." At the project's inception, you can prepare an outline of all final documentation and include this in your definition. Throughout the project, the responsible task manager or other specialist provides a few sentences or paragraphs when each key task is completed. These are inserted into the outline at the correct place. This is relatively painless and provides the essential information for final documentation later. These contributions might be provided via a local area network or magnetic media, thus reducing transcription requirements.

Final documentation can be aided by creeping documentation throughout the project.

COMPLETION CONSEQUENCES

Project completion may be viewed as a boon or doom. Senior management, the project manager, and the project personnel may not all see it the same way.

Three Affected Parties

For the project manager, completion may be an opportunity for promotion, but many project personnel may find themselves laid off if there is no other work. If the project was badly managed, its manager may receive a less favorable assignment in the future, and personnel who did outstanding jobs may have choice future assignments. The company's view of the new product development project's outcome normally depends almost entirely on the market's acceptance and the project's profitability.

Everyone does not have the same stake in completion, and the new product development project manager must understand the differences.

Thus, there is no reason to assume that all three parties will have the same view of project completion. Project managers must realize that they may have a very different stake in ending the project than the other two parties. Consequently, the project manager must prod, cajole, or offer inducements to those for whom completion is not obviously desirable.

Completion consequences are also influenced by the reasons for termination. It is certainly best to end the project because all the objectives have been satisfactorily achieved. It is a bad situation if one or more dimensions of the Triple Constraint have been missed substantially.

Personnel Reassignment

Project completion requires reassignment of people. We have now come full circle. The (temporary) project is no longer imposed on the rest of the (permanent) organization. This frequently necessitates a reorganization of the parent entity because the mix of work is such that the previously satisfactory organization is no longer appropriate.

The other crucial aspect of personnel reassignment is timing. If a person's next assignment is a choice one, he or she will normally be anxious to start and will lose interest in completing the present project. Conversely, if someone's next project assignment is undesirable, he or she may stall. When no assignment is obvious and layoff or termination is probable, personnel may even attempt sabotage to stretch out the present project assignment. Incentive bonuses for timely completion may be useful to counteract this potential problem.

A person's perception of what will happen when the project ends will affect his or her work as termination approaches.

Project managers can cope with these tendencies to some extent by selecting the time they inform project personnel of their next assignment. But if the company has a reputation for terminating personnel at the end of projects, there is little project managers can do. The best situation is one in which all project personnel can count on their good work being recognized and appreciated and there being a selection of future assignments.

Even if no specific new project assignment is available when personnel need reassignment, there are still options. For example, personnel can write an unsolicited proposal, prepare an article for publication, work on a feasibility effort, or attend a short course or seminar. Temporary assignments such as these can be used constructively to fill in valleys in the project work load. They also can be used as a motivational tool if they are authorized so as to make participation a mark of recognition for a job well done.

There are many options for personnel reassignment when projects end.

Organizational Changes Due to Completion

When any project ends, for whatever reason, the organization is altered. There are now new products that can be sold—perhaps new markets, customers, and users—and the promising prospect of still further future growth. There is a new body of knowledge in the organization. This is not just tangible information, but also new skills for many people. New working relationships have been established, both within and external to the organization, and these alter the informal organization, even if the formal organization remains unchanged.

INCREASING THE ODDS OF SUCCESS

Factors both external and internal to the multifunctional project team influence how well a project satisfies its Triple Constraint.

External Factors

Senior management that is responsible for or has cognizance over a new project development activity seems to be divided into two broad categories: knowledgeable and shortsighted. The shortsighted senior managers tend to emphasize the "we versus them" aspects and to some extent create an adversary relationship with the new product development project manager and the multifunctional project team.

There are two types of senior management: knowledgeable and shortsighted.

Knowledgeable senior managers realize that their stake in the new product project's success is ultimately just as great as that of the new product development project manager and the multifunctional project team. Thus, knowledgeable senior managers become involved in the new product development project in an effective and helpful, as opposed to destructive, manner. Such senior managers attend scheduled periodic and episodic (milestone) reviews at times convenient to the multifunctional project team. Beyond this, they attempt to ask the tough questions and carry out probing reviews of the team's work, not to embarrass but to assure that all significant issues have been dealt with appropriately. As one example, senior managers can assure that unanticipated resource requirements are satisfied promptly.

High priority projects inevitably seem to have better outcomes than lower priority projects because they tend to win all competitions for physical and human resources. This is not to say low-priority projects lack top management support; top management clearly wants all projects to succeed, but the lower priority projects are at a relative disadvantage.

Clear and stable project objectives are a *sine qua non* of project success. Objectives can and do change during the course of many projects, but not on a daily or hourly basis. Thus, committing these objectives to writing helps fix them in everyone's mind. Revising them when they must occasionally change is also a requirement of success.

Internal Factors

A qualified, experienced, competent leader is vital, as is a balanced team. Having a team with a balance of skills and getting teamwork from it can be somewhat contradictory. People with very similar backgrounds tend to get along better; so it is easier to promote teamwork in a group composed entirely of, for instance, electrical engineers. Nevertheless, a successful project usually requires a multifunctional project team be composed of more than electrical engineers. Thus, the project manager is confronted with merging people with diverse backgrounds into an effective and harmonious team.

A good leader, a balanced team, the right-sized work packages, careful replanning, and orderly termination contribute to project success.

Having the properly sized work packages helps you avoid two potential problems. Complex, difficult work packages should not be assigned to junior people, who may be overwhelmed by them. Simple work packages should not be assigned to senior people, who will not be challenged by them.

Because projects will almost never be carried out exactly in accordance with the original plan, replanning is a constant requirement in project management. Project termination, especially the reassignment of personnel, requires active planning well before scheduled completion.

CONTINUING SERVICE AND SUPPORT

Continuing service and support are normally an obligation. Therefore, it must be clear at the outset who is responsible for them and when. Because this is sometimes left to be decided when the project is nearly completed, when time is short and there is great pressure to meet the announced shipping date for the new product, it can leave a potential Pandora's box at the end of the project. Some of the issues that must be considered and resolved early—ideally, at project inception—are the responsibility for, resources dedicated to, and timing of the following:

It must be clear at the outset who is responsible for continuing service and support.

Continuing product support
    Customer service
        Telephone
        On-site
    User training
    Spare parts
    Maintenance
        Periodic
        Fire fighting
Follow-on product development
    Upgrades to the product
    Additional new product family members
Project audit (or post mortem)

The Project Audit

A project audit is important for three reasons:

1. To record lessons learned so as to do better in the future, which enables the company to improve continually
2. To provide immediate feedback to the next new product development project, whether it is an enchancement (for example, another member of the same product family) or a totally new effort
3. To maximize salvage value, if the project is terminated short of its planned objective

Therefore, regardless of whether the company's senior management conducts or requires an audit, every project manager should implement one.

An audit is an essential project activity.

    Depending on the specific project, audits may be called post mortems (a generic label), postcommercialization evaluations (new product development), or postimplementation reviews (software). The rationale is, "Those who cannot remember the past are condemned to repeat it."

The postproject audit must be objective and the auditor(s) very carefully selected. Auditors must be able to ask probing questions without appearing to be a threat and be able to communicate clearly. They must also be acceptable to the people they interview. This implies seniority, maturity, and competence as well as management support. Finally, senior management must clearly articulate its endorsement to avoid such audits being perceived as disruptive or witch hunts.

It is crucial that the auditor have had no stake or substantive involvement in the project being audited. Typically, an outside auditor is required in the case of a small company and may also be the best choice for a large one, although in the latter case, the auditor may come from a different unit of the company. Audits of very large projects may require a team of auditors.

The auditor(s) must be objective.

All audits should consider customer, prospect, and noncustomer viewpoints as well as insider's views. The audit should include interviews with each of the principal participants independently, and key observations should be recorded. If the project is expected to take longer than six months, the auditor should conduct interviews every three months to establish reliable baseline data while the participants' memories are still fresh. Notes on these interim interviews should be filed without comment until the end of the project unless information requiring immediate action is uncovered.

The audit concludes when the final and interim observations are distilled, summarized, and discussed with the entire group. Based upon these discussions, proposed changes in procedures should be recommended to appropriate management.

Prematurely terminated projects often have salvage value, providing an opportunity to sell off technology or related proprietary information (for example, primary market research findings). Thus, the people involved should prepare a final report. This record may provide the basis for saleable knowledge. Alternatively, if the project is restarted, the record can provide a jump start for the restored project.

Abandoned projects may have salvage value.

PEOPLE ISSUES

After being assured that all project personnel have been reassigned (or laid off or terminated if necessary), project managers have two other things to do. First, they should send personal letters of thanks, appreciation, or praise to project personnel. Second, they should send a brief wrap-up report to senior management. It is smart to cite your own successful performance in this.

TYPICAL PROBLEMS

Sometimes not only subordinate personnel but also the project manager must change during the project's life. The manager for the initial phases may be great at the inception but become stale with time or bored by routine wrap-up activities. The solution in this case is to change project managers, and both senior management and the project manager must be alert to this possibility, which then

(continued)

requires time for the replacement project manager to review (and perhaps revise) the plan so he or she is comfortable with it.

The press or excitement of new items to do often leads to the omission of some postcompletion activities. Organizations frequently never get around to performing a formal post mortem, so the same mistakes are repeated in subsequent projects. The solution is to recognize your responsibility in getting these things done.

HIGHLIGHTS

- Although the project starts with the definition activity and ends with the completion activity, project completion and market acceptance depend on agreements reached during the definition activity.

- Personnel needs may change throughout the project life cycle.

- It is best to end a project because all dimensions of the Triple Constraint have been satisfied.

- The project manager must realize that project completion may not be good for all involved parties and plan for an orderly end well in advance of its scheduled time.

- Some factors that contribute to the success of a new product development project are outside the project manager's control.

- Some documentation requirements are easily satisfied if prepared incrementally.

- Continuing service and support activities may lead to future business opportunities.

- Records must be carefully kept to ensure that there is an effective postcompletion audit.

- Managers should send letters of appreciation to project personnel and a wrap-up report to the boss.

CHAPTER 19

Where Do You Go from Here?

In this chapter, we wrap up our discussion of the new product development project management process with a summary of the key points you should remember as you put the tools and techniques discussed in this book into practice now or on your next project. Finally, we list some sources for continuing your development of skills.

SUMMARY

This section contains a skeletal overview of two aspects of the key topics we have covered. First, there are several elements of any new product development effort that the new product development project manager should be certain he or she keeps in mind:

Product business case
Product financial return
Customer-driven product specifications
A product family
Multifunctional project team
The company's new product development roadmap
Product stages
Process improvement feedback
Continuous improvement

Second, the following list summarizes the five managerial steps for successful new product development project management:

1. Define
   • Specific, measurable, attainable goals
   • Clear Triple Constraint priorities

2. Plan
   - Golden Rule
   - Three dimensions
       Work breakdown structure
       Time-based critical path schedule
       Task budgets
   - Contingency
   - Microcomputer software
3. Lead
   - Understand own behavior
   - Seek other people's strengths
   - Anticipate interpersonal conflict
   - Negotiate work packages
       Big WBS tasks for senior people, small ones for juniors
       Golden Rule
   - Communicate thoroughly
4. Monitor
   - Project reviews
       Get and understand reports
       Ask questions, especially nondirective ones
       Task complete or not
       Try to avoid impending problems
   - MBWA
   - Change resources
5. Complete
   - Unambiguous specification
   - Incremental documentation
   - Reassign and thank entire team
   - Carry out postcompletion work

When you start a new project, you should review these lists. Then turn to the chapter highlights, where appropriate, for a more detailed refresher. When necessary, you can then review portions of this book.

THE SPECIAL CASE OF NEW PRODUCT DEVELOPMENT FOR ANOTHER COMPANY

Many companies develop new products to be included in another company's new product. Examples include Connor, Quantum, and Seagate, which develop magnetic tape drives for computers, and Applied Magnetics, which develops magnetic heads used in the tape drives (and other applications). Other examples include the many companies that develop specialized parts such as galleys, seats, and audiovisual entertainment systems for commercial aircraft cabins or companies that develop portions of the airframe itself. This kind of new product development might be viewed narrowly as merely parts supply or subsystem production, but this overlooks the reality that each such product is unique and normally may not be furnished to another company. When you carry out this kind of new product development, there are many special issues with which your company must contend. An important element is the degree to which the product you are providing can be "productized" or offered to a broader market. This unique "sell-design-build" product development cycle calls for special vendor-customer relations. (The government contractor is a common occurrence of this unique product development situation.)

To begin with, your company probably will have to write a proposal to the firm that will be your company's customer, and you will have to compete to win the business. Your new product development schedule will be dictated by your customer's product introduction plans and schedule. In some cases, you may be brought in before the specification is fully complete, and you will really be working as a member of their multifunctional project team. Some of your company's personnel may even be located at your customer's company location. This situation is increasingly common (see Figure 1-1), because close teamwork with suppliers is frequently cited as one of the ways in which companies have shortened time to market.

Pricing such new product business is a major challenge. The production quantities of your product will depend on the sales of your customer's new product. The rate at which you must deliver may well differ from the forecast in the original request you receive. If your customer has a just-in-time (JIT) production system, your company may find itself holding inventory for its factory. Two issues you should consider very carefully are estimation inaccuracy and continuing service and support.

Estimation Inaccuracy

Several factors affect estimation accuracy. Of these, an imperfect definition of project scope is the most common. Either a sponsor or a performer may be the cause of the error, but it usually is attributable to both.

There may also be poor estimates of either time or cost. The rush to prepare a proposal and submit it in accordance with the bidding requirements may preclude there being sufficient time to do a good job of estimating. There is so much inherent uncertainty in some tasks of some projects that a poor estimate is almost a foregone conclusion.

Many jobs are proposed with deliberate underestimates of the amount of time or money it will take to perform them. This is the so-called "buy-in" situation in which a bidder attempts to win a job by making a low bid. This does not require an illegal misrepresentation, although that may be the case. It may result from a deliberate attempt to make optimistic assumptions about all the uncertainties in the proposed project as well as to omit all contingency from the estimates. In a sense, the bidder is making an estimate of time and cost that could occur perhaps 1 or 0.1 percent of the time rather than attempting to make an estimate near the mid range of possibilities.

Uncertainty and "buy-ins" can cause poor estimates.

Buy-in bids are much more prevalent where the contemplated contract will be a cost reimbursable form and the bidding contractor will not bear the financial burden of having made a low bid. They can also occur in a fixed price contract situation where the bidding contractor is confident that the customer will request changes in scope. Such changes will provide a "get well" opportunity: Increases in both time and cost for the main project can be added onto or concealed in renegotiations necessitated by changes of scope requested by the customer.

Continuing Service and Support

The contractor should view continuing service and support as an opportunity and not merely as an obligation. His or her employees will be working with the customer's personnel, providing continuing service and support, if it is included in the project. In so doing, they will have informal opportunities to explore ideas with the customer's personnel and hear about real problems the customer is facing. Thus, these contacts provide the basis for future business opportunities.

Some projects require postcompletion service and support.

SELF-TEST

As you contemplate becoming a new product development project manager or the boss of such a person, you might want to ask which of the following statements you would want a new product development project manager to make.

I know what to do; preparing a formal schedule plan is just a waste of my time.
Okay, here's my bar (Gantt) chart, which should satisfy your requirement.
We know what to do, and preparing a formal plan is just a waste of our project's time.
I'll make a schedule for my project.
We'll make a project schedule for the project and support tasks.
We keep our project's schedule up-to-date.
We constantly try to anticipate upcoming problems.
We normally develop alternative solutions to potential problems before they occur.
We routinely complete our projects on or ahead of time.
We always conduct a project completion audit.

OTHER SOURCES OF HELP

Sometimes you may still find that you are "in over your head." In such a case, you may wish to retain a management consultant for assistance. The organization that certifies individual management consultants is

The Institute of Management Consultants, Inc.
521 Fifth Avenue (35th floor)
New York, New York 10175-3598
Telephone: (212) 697-8262
Facsimile: (212) 949-6571

There are qualified management consultants unaffiliated with IMC. However, you have a better assurance of professional and ethical assistance if you choose a properly certified consultant.

Another source of help and an organization you should seriously consider joining is

Product Development and Management Association
c/o Thomas P. Hustad
Graduate School of Business
Indiana University
801 West Michigan Street
Indianapolis, Indiana 46202-5151
Telephone: (800) 232-5241 or (317) 274-0887
Facsimile: (317) 274-3312

PDMA is an international association designed to serve people with a professional interest in improving the management of product innovation. Its basic purpose is to seek improvement in the theory and practice of new product planning and development. PDMA has a unique diversity of membership that differentiates it from other professional organizations. Its membership includes practitioners, service providers, and academics from marketing, technical, and manufacturing disciplines. PDMA sponsors an annual international conference, numerous chapter meetings throughout the United States, and occasional short-stay conferences on special topics. PDMA also publishes the *Journal of Product Innovation Management,* which contains articles (both experiential and research), abstracts of pertinent other publications, and reviews of books of importance to practitioners.

CONTINUING PROJECT MANAGEMENT SKILL DEVELOPMENT

Reading this or any other book will not make you an expert new product development project manager. It takes time and practice for the skills to become second nature—and you will have to develop your own style, consistent with your skills, interests, and personality. There is no substitute for your own experience. Thus, you should continue to experiment, read, and seek out other sources for continuing education.

Reading

The citations at the end of this book can be used to identify other sources available at the time this book was published. More will be published in the future; so you should watch for these.

Continuing Education

Many universities and commercial organizations sponsor "live" training, such as seminars of one-day to one-week duration. These vary in quality, teaching method, and subject matter; so you should determine who will lecture and lead these training programs. Then try to check out references by talking with prior participants. We find one-day seminars superficial (with the privately sponsored, commercial versions consisting largely of a "hard sell" for the sponsors' publications). The five-day short course versions may require more time (and money) than is necessary, so the two- or three-day versions are most appropriate for your consideration.

The unique value of attending seminars or courses is the interaction with other new product development project managers. No amount of reading or passive observation of videocassette courses—even if the materials are outstanding—can give you real practice with the development of human relationship skills, which are crucial when you must work with people.

If you cannot attend a university or commercial seminar, you may want to explore having such a seminar conducted in your own organization. Although you will not obtain the stimulus of interacting with personnel from other, different organizations, you will have a seminar customized to your own specific situation (perhaps using your own forms as examples), and usually such an "in-house" seminar can be much more cost-effective than sending many people to other seminars.

A FINAL THOUGHT

Good luck! May all your new product development projects be successful.

Appendices

APPENDIX A

Abbreviations Used in New Product Development Project Management

ADM	Arrow Diagraming Method
AIN	Activity-in-Node
AOA	Activity-on-Arrow
AON	Activity-on-Node
ATI	After-Tax Income
BTI	Before-Tax Income
CDR	Critical Design Review
CE	Concurrent Engineering
COGS	Cost of Goods Sold
CPM	Critical Path Method
DCF	Discounted Cash Flow
DFMA	Design For Manufacturing and Assembly
DOE	Design of Experiments
EAC	Estimate at Completion
ECN	Engineering Change Notice
EF	Earliest Finish
EIN	Event-in-Node
ES	Earliest Start
E, S, & H	Environmental, Safety, and Health
ETC	Estimate to Complete
G & A	General and Administrative
HSP	Hoshin Strategic Planning

IMC	Institute of Management Consultants
IPD	Ingegrated Product Development
IRR	Internal Rate of Return
ISO	International Standards Organization
JIT	Just-In-Time
LAN	Local Area Network
LF	Latest Finish
LS	Latest Start
MBC	Management by Commitment
MBO	Management by Objectives
MBR	Management by Results
MBWA	Management by Walking Around
NPD	New Product Development
NPV	Net Present Value
OEM	Original Equipment Manufacturer
PBS	Project Breakdown Structure
PDM	Precedence Diagraming Method
PDMA	Product Development and Management Association
PDR	Preliminary Design Review
PERT	Program Evaluation and Review Technique
PGS	Productivity Gain Sharing
PIMS	Profit Impact of Marketing Strategies
PLC	Product Life Cycle
PM	Project (or Program) Management or Project (or Program) Manager
PO	Purchase Order
PR	Purchase Requisition
QA	Quality Assurance
QC	Quality Control
QFD	Quality Function Deployment
R, D, & E	Research, Development, and Engineering
ROA	Return on Assets
ROAE	Return on Assets Employed
ROE	Return on Equity
ROI	Return on Investment
ROTC	Return on Total Capital
SOW	Statement of Work
SPC	Statistical Process Control
TBAOA	Time-Based Activity-on-Arrow
TEI	Total Employee Involvement
TQM	Total Quality Management
VOC	Voice of the Customer
WBS	Work Breakdown Structure
WO	Work Order

APPENDIX B

Glossary of New Product Development Project Management Terms

Activity.   A single task within a project

Arrow diagraming method.   A type of network diagram in which the activities are labeled on the arrows

Bar chart.   A scheduling tool (also called a Gantt chart) in which the time span of each activity is shown as a horizontal line, the ends of which correspond to the start and finish of the activity as indicated by a date line at the bottom of the chart

Bottom up cost estimating.   The approach to making a cost estimate or plan in which detailed estimates are made for every task shown in the work breakdown structure and summed to provide a total cost estimate or plan for the project

Burst node.   In a network diagram, a node at which two or more activities commence after the completion of the preceding activity

Buy-in.   The process of making a cost bid in a proposal that is unduly optimistic or even actually less than the estimated costs for the project, which is done for the purpose of winning the job

Chart room.   A room filled with planning documents displayed as charts, typically hung on the walls of the room, used on large projects, and usually marked to indicate current status

Commitment.   An obligation to pay money at some future time, such as a purchase order or travel authorization, which represents a charge to a project budget even though not yet actually paid

Concurrent engineering.   The practice of developing a new product in which the design engineering work is done concurrently with the process or production engineering work, which is better practiced as integrated product development involving all functions (not merely the engineers from varied functions)

Contingency.   An amount of design margin, time, or money inserted into the corresponding plan as a safety factor to accommodate unexpected and presently unknown occurrences that judgment suggests will occur during the project (also called reserve)

Critical path.   In a network diagram, the longest path from start to finish or the path without any slack, thus, the path corresponding to the shortest time in which the project can be completed

Documentation.   Any kind of written report, including such items as final reports, spare parts lists, instruction manuals, test plans, and similar project information

Dummy activity.   An activity in a network diagram that requires no work, signifying a precedence condition only

Earliest finish.   In a network diagram schedule, the earliest time at which an activity can be completed

Earliest start.   In a network diagram schedule, the earliest time at which an activity can be started

Event.   The end of any activity or task

Float.   Same as slack time

Functional organization.   The form of organization in which all people with a particular kind of skill (such as engineering) are grouped in a common department, reporting to a single manager for that particular functional specialty

Hardware project.   A project in which the principal deliverable item is a product or functioning device of some sort

Hurdle rate.   The discount rate used by a company to calculate net present value in juding the attractiveness of any investment, including new product development

Integrated product development.   The practice of developing a new product in which a multifunctional team from, at least, the marketing, engineering (or equivalent), and production functions work concurrently

Internal rate of return.   That value of the discount rate for which the net present value is zero

Latest finish.   In a network diagram schedule, the latest time at which an activity can be finished

Latest start.   In a network diagram schedule, the latest time at which an activity can be started

Matrix organization.   The form of organization in which there is a project management functional specialty as well as other functional specialties and where the project management function has responsibility for accomplishing the project work by drawing upon the other functional specialties as required

Merge node.   In a network diagram, a node at which two or more activities precede the start of the subsequent activity

Milestone.   A major event in a project, typically one requiring the customer to approve further work

Net present value.   The monetary value of a discounted cash flow, typically calculated at a particular discount rate

Network diagram.   A scheduling tool in which activities or events are displayed as arrows and nodes in which the logical precedence conditions between the activities or events are shown

Periodic review.   Any kind of project review conducted on a periodic basis, most commonly a monthly project review

Phase.   That portion of the new product development process devoted to a given (sometimes specified) set of tasks, typically defined or established by a company as part of its new product development procedure

Planning matrix.   A matrix in which planned activities are listed on one side (usually the left) and involved people or groups are listed across a perpendicular side (usually the top) and where involvement of a particular individual or group in a particular activity is signified by a tic mark where the row and column intersect

Precedence diagraming method.   A type of network diagram in which the events and activities are labeled in the nodes or boxes

Program.   Used interchangeably with "project," as in "program management" or "program manager"

Program Evaluation and Review Technique (PERT).   The form of network diagram in which events are displayed as nodes and connecting arrows indicate the precedence constraints

Project.   An organized undertaking utilizing human and physical resources, done once, to accomplish a specific goal, which is normally defined by a Triple Constraint

Project cost accounting system.   A cost accounting system that accumulates actual costs for projects in such a way that total costs for all work in an organization can be allocated to the appropriate projects, normally providing monthly cost summaries; also used in cost planning to summarize the detailed task cost estimates

Project organization.   The form of organization in which all or nearly all the people working on a project report to the project manager

Project plan.   The entire plan for a project, consisting of the work breakdown structure, network diagram, and task budgets, but sometimes taken to mean only the network diagram

Project team.   A term used in this book to designate the personnel working on a project who report to the project manager administratively, not merely for the work on the project

Quality Function Deployment (QFD).   A system for translating market research representing the voice of the customer into the design and production requirements for products and services

Roadmap.   A picture or summary graphic portrayal of the main elements of a company's new product development procedure

Simultaneous engineering.   Same as concurrent engineering

Slack time.   In a network diagram, the amount of time on any path other than the critical path that is the difference between the time to a common node on the critical path and the other path

Software project.   A project in which the principal deliverable item is a report or other form of documentation, such as a computer program

Stage.   Same as phase, although frequently used in "stage gate" to indicate that each phase or stage ends with a "gate" or management approval to enter the next stage

Statement of work (SOW).   That statement of exactly what will be delivered and when

Subcontractor.   An organization, usually a company, working for another organization on some aspect of the project for which the other organization is under contract

Support team.   A term used in this book to designate the personnel working on a project who do not report to the project manager administratively

Task.   A small part of a project

Task force.   An ad hoc group designated to cope with a project, similar to a project organization although frequently staffed with personnel on part-time assignment, usually adopted by a functional organization having only one project or at most a few projects at any given time

Time compression.   The act of reducing the planned time for an activity, accomplished perhaps by adding unplanned staff or using overtime

Top down cost estimating.   The approach to making a cost estimate or plan in which judgment and experience are used to arrive at an overall amount, usually done by an experienced manager making a subjective comparison of the project to previous, similar projects

Topical review.   Any kind of project review devoted to a single topic, such as a final design review or a manufacturing review

Total Quality Management (TQM).   A philosophy addressing the quality of management rather than the management of quality; basic tenets include a focus on the customer and continuous improvement of the process

Triple Constraint.   The term used in this book to describe the three key project objectives that must be simultaneously accomplished—the performance specification, the time schedule, and the monetary budget

Venture organization.   The form of organization used in some large organizations where a three- or four-person team, itself functionally organized, is established within the larger organization to develop and commercialize a new product

Work breakdown structure (WBS).   A family tree, usually product oriented, that organizes, defines, and graphically displays the hardware, software, services, and other work tasks necessary to accomplish the project objectives

APPENDIX C

Planning Checklists for New Product Project Managers

SOME NEW PRODUCT DEVELOPMENT TASKS FOR YOUR CHECKLIST

This list is intended to be only a guide to increase the likelihood that you will not omit some required task from your critical path network schedule. No single new product development project will require all of these tasks to be completed. Conversely, the development of almost any new product or service will require some additional tasks that are not included in this list. Your business may require that some of these occur either earlier or later in the development activity; nevertheless, you can use this list as a starting point on your next new product development program. Then, as you modify the list, you will be better prepared for subsequent new product development efforts.

Idea For	*Feasibility*
unsolved market problem	initial screening
new use for old product	fit with strategy
new technology to exploit	fit with culture
unique product	barriers to entry
new process	for us

Feasibility (cont.)
    for competition
    patent search
  markets
  technology
  manufacturing
  service and support
  timing
impact on other products
preliminary market assessment
preliminary technical assessment
preliminary manufacturing assessment
preliminary business assessment
establishment of proprietary confidentiality policy
product concept
concept testing
feasibility demonstration
breadboards(s)
computer simulation
bench chemistry
design analyses

Optimization

formation of triad team
market research
  secondary
  primary
  names of first three to six buyers
  ability of customer to assimilate product
  competitive analysis (in-kind and functional)
  standards
  legal and regulatory constraints
  political constraints
  social acceptability
market segmentation
product positioning
plan to phase out replaced products
trade-in plan
distribution plan
project schedule and budget plan
  resource availability

Optimization (cont.)
  facility (for example, skunk works) or team location
  priority (compared to alternatives)
assessment of liability risks
business analysis
specifications
  concept sketches
  product attributes: musts and wants
    technical characteristics
    user interfaces
    target introduction date
    styles (including shapes, colors, textures)
    sensory features
    sizes
    accessories
    consumables
    adjunct services
    safety
    warranties
  factory cost or sell price and gross margin
  follow-on products and services

Design

system engineering
block level design
design and concept reviews
failure analysis
power consumption and dissipation analysis
industrial design and styling
service access and diagnostics
user controls and interface
design drawings
parts list
accessories
consumables
bill of material
software walk-throughs
sources of supply
  long lead time items

Design (cont.)

quantities versus time
inventory plan
component test plan
environmental risks and constraints
product test procedures
manufacturing of test equipment
building of prototype(s)
prototype testing
  device characterization
  qualification life tests
  endurance tests
  environmental tests
  reliability tests
  maintainability demonstration
quality assurance plan
in-house product testing
design revisions
product name
public relations plan
product packaging
determination of environmental
  limits
shelf life tests
regulatory approvals
  underwriters laboratories
  Federal Communications
    Commission
  Federal Aviation Administration
  Food and Drug Administration
  Environmental Protection Admini-
    stration
  premarket notifications
  material safety data sheets
  Toxic Substances Control Act
customer tests of product
labeling requirements
shipping container design
international distribution plan
financial evaluation
obtaining of protection
  patents
  copyrights
  trademarks

Preproduction

process development
soft tooling and pilot run
training of manufacturing personnel
design verification test
reliability verification test
manufacturing verification test
quality verification test
software validation test
documentation releases
materials requirements plan
configuration releases
trial production
locating second sources of supply
test marketing
beta test sites
pilot plant
assembly fixtures
test fixtures
production cost estimates
precommercialization business analysis
tooling
export license

Production

production start-up
sales demonstration samples
training of
  trainers
  salespeople
  product installation teams
  product support specialists
  service personnel
  users
training materials
  demonstration units
  transparencies
  programmed instruction
  videocassettes
  software data sets
  accessories
  consumables
  handouts

Production (cont.)
semicommercial production
advertising campaign (all markets)
  key account preannouncement
  customer
  suppliers
  manufacturers of congruent equip-
    ment and supplies
  trade
  trade shows
  influencers
  financial community
  end user
  catalog sheets
  coupons
  package artwork
  product tie-ins
  news releases
  mailing labels
literature (all languages)
  installation instructions
  data sheets
  brochures
  warranty cards
  user instructions
  training manuals
  service instructions
sales tools
  terms and conditions
  trade-in policy
  adjunct services, accessories, supplies,
    and products
  acceptance test procedure
  giveaways
  contests
  point-of-sale displays
  testimonials
  discount schedule
  sales quotas
  user groups

Sales

market launch
  regional

Sales (cont.)
  national
  multinational
  global
full-scale production
service spares
field sales
after-sales support
brand building
market growth

Postintroduction Appraisal

post mortem
lessons learned
revision of new product development
  procedures

*Hardware Engineering or Study
Projects*

system review
system approval
system test criteria
system test plan
detailed hardware specification
customer-furnished equipment
power requirement plan
weight control plan
breadboard design
breadboard fabrication
breadboard test
block diagram
schematic diagrams
circuit diagrams
conceptual design review
preliminary design review
critical design review
final design review
prototype design
prototype fabrication
prototype test
design freeze
drawings freeze
functional designs

*Hardware Engineering or Study
    Projects (cont.)*

system logic design
optical design
mechanical design
electronic design
thermal design
subsystem hardware implementation
subsystem software implementation
subsystem integration
subsystem review
subsystem approval
subsystem test criteria
subsystem test plan
make/buy decisions
long lead items
special test equipment
commercial test equipment
calibration of test equipment
software tests
data reduction plan
operational software
subsystem cabling
system cabling
installation planning
experimental development plan
support plans
support instrumentation
facilities
training plans
repair facilities and requirements
inspection
preshipment review
customer inspection
customer acceptance
preparation for shipment
shipment
customer support
qualification test
flight acceptance test
launch support
mission support
personnel recruitment
personnel reassignment

*Hardware Engineering or Study
    Projects (cont.)*

documentation
  project plan
  integrated schedule
  functional requirements document
  environmental requirements document
  environmental test specifications
  environmental test procedures
  environmental test reports
  interface control
  safety plan
  configuration control plan
  failure mode and effect analysis
  reliability and quality assurance plan
  development test plan
  acceptance test procedure
  calibration plan
  ground data-handling plan
  experiment development plan
  expendables consumption
  engineering drawings and drawing
    list
  parts list
  electronic parts acquisition and screen-
    ing plan
  materials documentation
  manufacturing release
  periodic reports (for example, monthly)
  special reports
  final reports
  instruction manuals
reviews
  manufacturing review
  management review
  critical design review
  preshipment review
  internal project review
  subcontractor progress review
  customer review

Programming Projects

applications requirements
systems requirements

Programming Projects (cont.)
system inputs
system outputs
detailed architectural design
design specifications
functional specifications
security plan
system test and acceptance specifications
feasibility studies
file and data requirements
cost/benefit analysis
system design
program design
system conversion plan
shipment and delivery
turnover to operations
postimplementation reviews
supplies
training aids

Programming Projects (cont.)
review procedures
hardware requirements
personnel capabilities
milestone reviews
milestone documents
product specifications
project plan
operator instructions
user instructions
library
project index
change control system
data base administration
technical interface manuals
hardware reference manuals
release information
systems reference manual

Appendix D

Financial Analysis

You must be able to demonstrate that a new product development program will be profitable. Internal rate of return and net present value are straightforward ways of analyzing profit potential and can be used to analyze other prospective investments. Five estimated quantities determine both the internal rate of return and net present value. Each financial measure has its own sensitivity to changes in the five estimated quantities.

PROGRAM FINANCIAL JUSTIFICATION

Consider the case of the computerized axial tomographic (CAT) scanner, developed by an English company, EMI. EMI's technical director, Godfrey Hounsfield, received the joint Nobel prize for medicine for developing the CAT scanner. However, two press reports reveal a less satisfactory business outcome:

> Rarely can a Nobel prize for scientific achievement have been followed so closely by a booby prize for business failure as in the case of the CAT scanner business of Britain's EMI.[1]

> Although the scanner was one of the outstanding advances in medical sciences in recent decades, EMI was unable to make it a financial success.[2]

A new product development program must have a prospective financial return commensurate with its inherent risk. Just as you would try to sell a product to a Frenchman by speaking French, you must justify new product development programs to financially oriented top management or investors. Consequently, you must translate the new product development project's justifications into financial terms. Winning the Nobel prize, despite the attendant prestige, just is not good enough, nor is distribution into an exotic and attractive foreign country or the requirement to use the latest computer-controlled production machinery. No matter how exciting the market, technology, or manufacturing aspects, the effort has to make money for the company.

Suppose you can get $6 for selling a product that costs you only $5 to produce and sell. That is a pretax profit of $1 per $6 of sales, or about 8 percent after-tax profit on sales. That profit rate exceeds the earning rate of the vast majority of American companies. Would you like to propose such a new product development program to your management? As you will see shortly, profit rate is only one financial issue you must consider.

Although different companies use different financial justification techniques, most now use the internal rate of return (IRR) or the net present value (NPV) to evaluate major investments. Thus, if new product development programs can be described in terms of IRR or NPV, these programs and their approvals are put on the same footing as capital investment approvals, for instance, whether to build a new warehouse, capital asset procurements, manufacturing capacity expansion, and similar undertakings, allowing you to consider all investment opportunities on a comparable risk-adjusted basis. When presented this way, new product development programs frequently will show very attractive returns. But if these programs do not show attractive returns or provide a high assurance of an intangible competitive advantage (which may not be easy to analyze quantitatively), perhaps they should not be funded.

Microcomputer spreadsheet software is ideally suited to calculate IRR and NPV, as well as other common financial measures. Assuming basic data are available, use of almost any spreadsheet program is quick and easy. (Other specialized software is also available, but its use is not required.) Using such software will help you identify any critical areas of the new product development project requiring more detailed examination. Because a financial model summarizes the involvement of all departments, you can use it to aid communication with the other functional managers involved in the project.

If you do not know what method of financial analysis your company now uses, find out. This is crucial for two reasons. First, you want to be sure any new product program you advocate is well justified by your company's own standards. Second, you want to know enough either to do the analysis yourself or to be able to check the reasonableness of someone else's calculations.

INTERNAL RATE OF RETURN

IRR and NPV are two of several financial measures corporations use to analyze the attractiveness of prospective investments. Profit margin, payback period, and return on investment (ROI) are some others, but each has substantial drawbacks. Profit margin is simple but does not contain any information about the investment required to obtain the profit margin, nor does it provide any measure of risk. Although payback period (Figure D-1) ignores everything that occurs after payback, it is a simple and useful indicator of risk because a long payback period is riskier than a short one. ROI is certainly an important calculation (Figure D-2). However, for future investments, it can be done only on a period-by-period basis (for example, fiscal year), or it must be

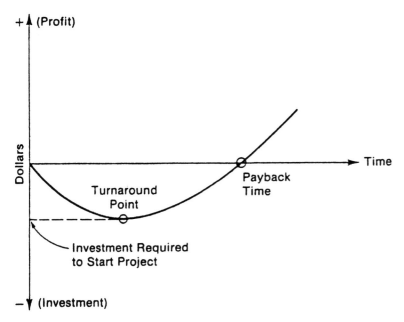

FIGURE D-1.   Payback period.

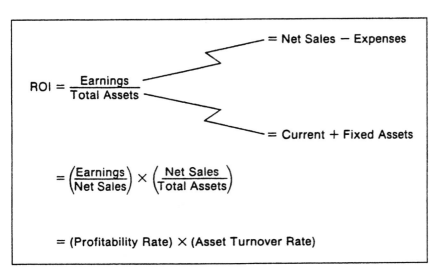

FIGURE D-2.   Return on investment.

averaged over some longer time period. ROI is also a generic label for any of several measures:

- Return on investment

$$\text{ROI} = \frac{\text{net income (after tax)}}{\text{shareholders' equity} + \text{debt}}$$

- Return on equity

$$\text{ROE} = \frac{\text{N.I.}}{\text{S.E.}}$$

- Return on total capital

$$\text{ROTC} = \frac{\text{N.I.} + \text{after-tax interest on debt}}{\text{S.E.} + \text{debt}}$$

- Return on assets

$$\text{ROA} = \frac{\text{N.I.}}{\text{total assets (depreciated) on balance sheet}}$$

- Return on assets employed

$$\text{ROAE} = \frac{\text{N.I.}}{\text{total assets} + \text{accumulated depreciation} + \text{value of leased facilities}}$$

So it can be ambiguous. You should understand precisely which return measure(s) your own company uses. In common with the two preceding measures, ROI fails to account for the time value of money. NPV, which does reflect the time value of money, is related to IRR, but NPV requires that a discount (or hurdle) rate be preselected. Both NPV and IRR are forms of discounted cash flows (DCFs), in which money in future years is taken to have less value than today's money. Each of these techniques, and many others, is treated in financial texts in more detail.

IRR depends on the timing of expenses and receipts, thus recognizing the time value of money. Fortuitously, calculation of IRR provides all the information to permit quick determination of the other measures. The inherent assumption in IRR is that recovered funding is reinvested into the same venture, at the venture's rate of return. IRR is the rate by which the cash flow must be discounted, such that the total discounted cash flow is equal to zero over the period representing the life of the investment, for instance, ten (or seven, five, or any other appropriate duration of) years. Expressed another way, IRR is the discount rate for which the program's NPV is zero. Thus, IRR provides a direct comparison against the cost of money.

CALCULATING IRR

An illustration of a DCF will provide you with a simple tool for daily use.

Required Data

To calculate IRR or NPV, you must know the total cash flow for each year of the program (or venture or project or development). To determine cash flow, you must make five assumptions or estimates for each year:

1. Company sales resulting from the development program
2. Manufacturing costs attendant to the sales
3. Development expense to achieve the sales
4. Operating costs attendant to the sales
5. Capital expenditures to permit production, distribution, and so forth

It may seem like a lot of work to obtain this information, but it simplifies justification. It is prudent to make these estimates before engaging in a program rather than persisting in an unprofitable program. If you lack the information to make these five estimates, that is a clear danger signal that your program's ultimate success depends on unknowns. Therefore, the work required to make the estimates is an essential ingredient of a successful program. Because each of the five quantities is an estimate, the derived IRR is an estimate. Nevertheless, it takes only a few minutes to analyze a proposed development program's IRR or NPV and its sensitivity to the assumptions. In doing this analysis, you will gain a great deal of insight.

It is conventional to enter the five estimated quantities in the first five lines. Everything else depends solely on what you enter here. (In some cases, two other lines precede the top line. These two lines are unit sales and average sales price. Annual company sales are then calculated as the product of these two estimates. Similar elaborations are possible for other estimated quantities.) The numbers entered in lines 1 to 5 of Figure D-3 are an example of the kind of estimates (in thousands or millions of dollars) that might be made. Company sales (line 1) should be estimated by the marketing department or be derived from market research, and the estimate must reflect market reality, not wishful thinking. Criteria by which sales estimates may be made include the value of the proposed new product substitution to the customer, the sales volume of similar products, market share estimates where market size is known, or the number of units to be sold multiplied by price per unit. Use market research to improve these estimates whenever possible. It is essential that sales volume be consistent with manufacturing and distributing capacity.

The manufacturing and technical development people should estimate manufacturing cost (line 2). They can use the cost of similar products, a component parts list cost multiplied by a suitable markup factor, or, ideally, a detailed product breakdown and manufacturing plan. When the manufacturing rate will be nonuniform, it may also be helpful to use an order-mix simulation.

YEAR =	1	2	3	4	5	6	7	8	9	10
1 Company Sales			2.00	2.40	2.88	3.46	4.15	4.98	4.98	4.15
2 Manufacturing Cost			.80	.96	1.15	1.38	1.66	1.99	1.99	1.66
3 Development Expense	.50	.50								
4 Operating Expense			.40	.48	.58	.69	.83	1.00	1.00	.83
5 Capital Expense	1.00									
6 Depreciation	.20	.20	.20	.20	.20	.00	.00	.00	.00	.00
7 Gross Profit	.00	.00	1.20	1.44	1.73	2.07	2.49	2.99	2.99	2.49
8 Before Tax Income	-.70	-.70	.60	.76	.95	1.38	1.66	1.99	1.99	1.66
9 Income Tax	-.35	-.35	.30	.38	.48	.69	.83	1.00	1.00	.83
10 Net Income	-.35	-.35	.30	.38	.48	.69	.83	1.00	1.00	.83
11 Operating Cash Flow	-.15	-.15	.50	.58	.68	.69	.83	1.00	1.00	.83
12 Working Cash Required	.00	.00	.60	.12	.14	.17	.21	.25	.00	-.25
13 Total Cash Flow	-1.15	-.15	-.10	.46	.53	.52	.62	.75	1.00	2.32
14 Cumulative Cash Flow	-1.15	-1.30	-1.40	-.94	-.41	.11	.73	1.48	2.47	4.80

Internal Rate of Return (%) =   26.36

Discounted Cash Flows:
NPV @ 10%	1.85
NPV @ 15%	1.04
NPV @ 20%	.48
NPV @ 25%	.09
NPV @ 30%	-.20
NPV @ 35%	-.41
NPV @ 40%	-.57

FIGURE D-3.   Basecase spreadsheet example.

Development expense (line 3) may occur in marketing, manufacturing, certainly will occur in R, D, & E, and perhaps will occur in other departments; so each concerned department must provide appropriate estimates, ideally from detailed plans, including contingency.

Operating expense (line 4) is usually related to sales and might typically be some percentage of sales, say 25 percent or 30 percent. Each concerned department should provide these estimates from detailed plans.

Each department concerned in the development must estimate capital expense (line 5). This should include transportation and installation costs when applicable, as well as capital contingency. The estimates and the assumptions upon which they are based are crucial in determining the calculated quantities.

Determining Total Cash Flow

The other eight quantities (lines 6 through 13) are merely arithmetic manipulations of the five estimated quantities for each year of the analysis, typically ten years. The illustrative spreadsheet has been set up so that once the five estimated quantities are entered, it is very easy to calculate the resulting cash flow. The following specific calculations are required:

Depreciation (line 6) should be taken in accordance with current tax laws or, for simplicity of the illustration, as 20 percent per year (that is, straight-line depreciation for a five-year life). In general, companies prefer short (and accelerated) depreciation schedules because these will improve the IRR, and they will select these if available.

Gross profit (line 7) is merely the difference between company sales (line 1) and manufacturing cost (line 2).

Before-tax income (line 8) is the gross profit (line 7) less depreciation (line 6), development expense (line 3), and operating expense (line 4).

Income tax (line 9) can be calculated at the company income tax rate or, as in the illustration, assumed to be 50 percent of the before-tax income.

Net (or after-tax) income (line 10) is the before-tax income (line 8) less income tax (line 9).

Operating cash flow (line 11) is net income (line 10) plus depreciation (line 6). Because your company does not actually spend current year depreciation, this line is the cash available for use on an operating basis (that is, without consideration for any capital expenditures).

The working cash required (line 12) is the funding it takes to support a growing business to finance inventory and accounts receivable. Company practice may provide a guide for what this should be, although it is probably 25 percent to 35 percent of the sales increase of the current year compared to the previous year. The illustration is constructed on the assumption of 30 percent. The amount of line 12 can be calculated exactly for any given situation; the illustrative case demonstrates a simple way to insert a reasonable estimate of the cash required. As a general rule, (1) rapid sales increases, (2) low profits, and (3) long periods between incurring costs and getting paid all increase the working cash required. The reverse lowers it.

To get a better feeling for the working cash requirement, think of it as the money required to pay your bills prior to being paid by your customers. Consider the case in which you get paid $6 for something that costs you $5 to make and sell, which provides a nice profit. Assume you must pay the $5 of cost (for example, materials purchased from suppliers and wages for your workers) three months before you are repaid $6 by your customer. Assume your sales, and therefore your costs, increase 10 percent per month. This profitable growth situation (Table D-1) continues to require a cash investment as long as the sales continue to increase, even when money begins to flow in from customers. You use this cash to finance your inventory (raw material, work in process, and finished goods) and accounts receivable. You can easily see that the cash position at the end of the fourth month and later would be worse if the customer paid later, if profits were less, or if the rate of sales increase was more rapid. When sales flatten out and decline, the venture begins to generate cash, and line 12 on the spreadsheet becomes negative (because a *negative* working cash requirement is a *positive* cash flow).

TABLE D-1.   The Requirement for Working Cash

Month / Sale	1	2	3	4	5	6	7	8	9
First	<5>			6					
Second		<5.5>			6.6				
Third			<6.05>			7.26			
Fourth				<6.66>			7.99		
Fifth					<7.32>			8.78	
Sixth						<8.05>			9.66
Etc.									
Net Cash for Month	<5>	<5.5>	<6.05>	<0.66>	<0.72>	<0.79>			

TABLE D-2.   Cash Flow and After-tax Income

		Cash Flow	
After-Tax Income		No Capital Investment	With Capital Investment
Sales	1,000	1,000	1,000
Manufacturing cost	−500	−500	−500
Gross margin	500	500	500
Operating expense	−300	−300	−300
Depreciation	− 50		
Before-tax income	150		
Income tax	− 75	− 75	− 75
After-tax income	75		
Capital investment			−500
	75	125	−375
		Difference is depreciation	Difference is capital investment

The total cash flow (line 13) is the operating cash flow (line 11) minus the working
cash required (if it is positive). This is the predicted amount of actual cash required
(if line 13 is negative) or generated (if line 13 is positive) each year. This is not the
same as after-tax income, as illustrated in Table D-2. Total cash flow differs from
after-tax income because of capital expenditures (a cash flow outlay that cannot be
expensed in determining taxable income), depreciation (the noncash "expense"

allowed as a tax deduction for prior capital expenditures), and working cash required (the money it takes to finance or run the venture).

In the last year of the financial analysis (whatever year it may be), it is conventional to assume that the venture is liquidated, which means the total cash flow for that year is the sum of what is calculated normally plus the summation of all the working cash required to date. This is just another way of saying that the assets on the balance sheet are being sold off in the final year. This adjustment is noted at the end of line 13 of the spreadsheet in Figure D-3.

Payback Period

Having the total cash flow (line 13), you can easily determine the payback period (see Figure A-1) by looking at the cumulative cash flow and determining in which year it turns, and remains, positive. Line 14 is included in the spreadsheet illustration to demonstrate this.

Calculating the IRR or NPV

IRR may be calculated now in two ways from the total cash flow. First, most spreadsheet programs have an IRR function, and you may use this. Alternatively, you can calculate a few NPVs by using several discount factors. The NPV that is equal to zero provides the discount percentage that is the value of IRR. If your spreadsheet does not have an IRR function, calculate a few NPVs and estimate (or interpolate) to determine the IRR.

There are some cautions to observe in this method of program financial analysis. If the program goal is to produce a replacement product, you must subtract the profits the existing product would have produced had it not been replaced in the appropriate years. Similarly, if the new product is a member of a planned family, one product that is important for defensive competitive reasons because it fills a minor market niche can be marginal financially if the entire family taken as a whole is profitable.

Significance of IRR

What is the significance of an IRR value, say the 26.36 percent in the base case example (Figure D-3)? At the simplest, the IRR value of alternative investments can be compared, and resources can then be committed to the most promising undertaking(s).

For example, assume you can obtain an 8 percent (before-tax) yield on funds invested in some security issued by the United States government, which is the safest investment you can find. This is equivalent to a 4 percent after-tax return (at a 50 percent income tax rate), a value that can be compared to the IRR (which is also an after-tax value). In this case, we can say that there is a 100 percent certainty that a U.S. security investment will return 4 percent after taxes. As long as we have more than a 15 percent confidence that the alternative new product investment will go according to plan, we would favor it (because 0.15×26.36 percent = 4 percent, which is the same return as for the U.S. security). To put it another way, the risk factor in a venture must discount the IRR to allow comparison of investment opportunities.

Your company's basic strategy, and the reasons you have for undertaking new product development, as discussed in Chapter 1, should establish your minimum IRR target. As a general rule, forecasted new product development IRR values should exceed 20 percent, or perhaps 30 percent, because of the risk. If the forecasted IRR value is lower, there probably is a better investment available elsewhere.

There is a general caution about IRR and NPV or any other financial analysis. Such analyses involve assumptions such as the estimates on lines 1-5, the applicable tax rate, and the required working cash. Consequently, the actual return from the investment may differ from the calculated return. Thus, sometimes you should still undertake a new product development program or make a capital investment when the calculated IRR is low. This may be the lesser of evils in some situations. For instance, developing a new attachment to your existing product might have a marginal financial return, but this development could prevent a competitor from gaining access to your customer base.

There frequently are intangible factors not amenable to analysis. Consider a decision whether to make a capital investment in new automated production machinery. You can estimate the value of the labor saved precisely, but it is harder to assess the value of being able to produce early product models rapidly for examination by the multi-functional team, if you can quantify it at all.

Therefore, you should use IRR (or NPV) as a guide rather than an absolute criteria. If a thoroughly analyzed new product program has a high IRR (for instance, greater than 20 percent), it is probably a good undertaking, even though it should still be judged against other investment opportunities. If the calculated IRR is low (for instance, less than 10 percent), look around for a better new product investment opportunity, or, if you go ahead, understand the financial risk you are taking. In between these extremes, you must recheck assumptions and not just change the numbers because you do not like the calculated IRR. Do not, however, go ahead with any new product development program that has a low IRR just because someone in R, D, & E or marketing says every new product development is inherently risky. With careful screening and thoughtful analysis, you can improve your success rate. Finally, if you do not have enough information to estimate the IRR, that is a warning that you do not yet know enough about the proposed development program.

COMPARISON OF FINANCIAL MEASURES

Each financial measure has a different sensitivity to any change in the five estimated quantities. To illustrate this, consider four alternatives for the investments being made during the first two years of the base case (Tables D-3 and D-4). IRR and NPV yield the same relative ranking (that is, case C is most attractive, followed by case D, the base case, case B, and case A). Return on investment is not sensitive in the same way. This anomaly arises because ROI is not dependent on noncapital expenditures, namely, the initial development expense (line 3). ROI is thus undesirable for evaluating new product development programs. The discretionary investment you, as a manager, must make involves both expensed and capitalized items, and you want to be able to measure the return on the combination.

No isolated financial measure can be definitive. For instance, IRR does not reflect

TABLE D-3.   Estimated Quantities That Are Varied
as Input to Figure D-3 Obtain the Results
Shown in Table D-4

	Line 3 Development Expense		Line 5 Capital Expenditures
Year	1	2	1
Base case	0.5	0.5	1.0
Case A	0.5	0.5	2.0
Case B	1.0	1.0	1.0
Case C	0.5	0.5	0.5
Case D	0.25	0.25	1.0

TABLE D-4.   Comparison of Several Financial Measures (All cases have same sales, manufacturing cost, and operating cost in years 3 to 10.)

	Base	A	B	C	D
Years 1 & 2					
Total "start-up" charges	2	3	3	1.5	1.5
Year 5					
Gross margin % (7 ÷ 1)	60	60	60	60	60
Profit rate % (10 ÷ 1)	16.7	13.2	16.7	18.4	16.7
ROI % (10 ÷ Σ5)	48	14	48	106	48
Entire program					
NPV at discount factor = 20%	0.48	−0.16	0.03	0.81	0.71
Maximum negative cumulative cash flow (in year 3)	−1.40	−2.10	−1.90	−1.05	−1.15
Payback period (approximate years)	5.8	6.6	6.6	5.3	5.3
IRR%	26.36	18.49	20.29	33.29	30.42

the absolute size of the investment. For instance, a program with an IRR of 25 percent that has sales of $100 million might be better then an alternative with an IRR of 30 percent if the latter had sales in the same time period of only $10 million. Thus, although IRR is an excellent way to assess your new product development program, you should still use judgment and not rely entirely on myopic numerical analysis.

IRR AND NPV SENSITIVITY TO ESTIMATE ASSUMPTIONS

As noted, the five estimated quantities completely determine the resultant financial measures. Therefore, it is important to ascertain the sensitivity of IRR or NPV (or any other financial measure) to the assumptions. You can do so quickly by altering each

TABLE D-5.   IRR and NPV Sensitivity to Estimates

Case	IRR %	NPV (at 20% discount)
Base case	26.36	0.48
−10% development expense	27.10	0.53
−10% capital expense	27.50	0.55
−10% operating expense	27.63	0.59
−10% manufacturing cost	28.88	0.69
+10% sales	31.67	0.95

YEAR =	1	2	3	4	5	6	7	8	9	10
1 Company Sales			2.20	2.64	3.17	3.80	4.56	5.47	5.47	4.56
2 Manufacturing Cost			.80	.96	1.15	1.38	1.66	1.99	1.99	1.66
3 Development Expense	.50	.50								
4 Operating Expense			.40	.48	.58	.69	.83	1.00	1.00	.83
5 Capital Expense	1.00									
6 Depreciation	.20	.20	.20	.20	.20	.00	.00	.00	.00	.00
7 Gross Profit	.00	.00	1.40	1.68	2.02	2.42	2.90	3.48	3.48	2.90
8 Before Tax Income	-.70	-.70	.80	1.00	1.24	1.73	2.07	2.49	2.49	2.07
9 Income Tax	-.35	-.35	.40	.50	.62	.86	1.04	1.24	1.24	1.04
10 Net Income	-.35	-.35	.40	.50	.62	.86	1.04	1.24	1.24	1.04
11 Operating Cash Flow	-.15	-.15	.60	.70	.82	.86	1.04	1.24	1.24	1.04
12 Working Cash Required	.00	.00	.66	.13	.16	.19	.23	.27	.00	-.27
13 Total Cash Flow	-1.15	-.15	-.06	.57	.66	.67	.81	.97	1.24	2.68
14 Cumulative Cash Flow	-1.15	-1.30	-1.36	-.79	-.13	.54	1.35	2.32	3.57	6.25

```
Internal Rate of Return (%) =    31.67


Discounted Cash Flows:
    NPV a 10%        2.63
    NPV a 15%        1.64
    NPV a 20%         .95
    NPV a 25%         .46
    NPV a 30%         .10
    NPV a 35%        -.17
    NPV a 40%        -.37
```

FIGURE D-4.   Spreadsheet for case when sales are 10 percent higher.

assumed quantity and calculating a new IRR or NPV. For instance, each estimated quantity can be improved by 10 percent of the assumed value for all years being estimated and the resultant IRR or NPV calculated. In this instance, the sales and manufacturing cost estimates are most critical (Table D-5; Figure D-4 is the spreadsheet for sales increased by 10 percent, without any change in other estimates). Thus, in a real situation, you might choose to examine these in more detail. In general, IRR and NPV will always be most sensitive to sales. However, the relative ranking of the other four estimated quantities will vary, depending on the specific numbers.

IRR AS A MANAGEMENT TOOL

Consider the following hypothetical situation. You are the manager of a new product development program. At the end of the first year of the program, a revolting development becomes apparent: Design complexity causes manufacturing costs 25 percent higher than the original plan, namely, 50 percent of sales rather than 40 percent. The multifunctional team proposes the following three alternatives:

1. (X) Continue on the current schedule, but settle for manufacturing costs that are 50 percent of the sales price rather than the previously estimated 40 percent.
2. (Y) Spend an extra year and $0.5 million in development to reduce the manufacturing cost to the original plan, but accept a one-year delay.
3. (Z) Spend an extra $0.7 million in R & D development expense in the next year to recover the original plan.

The IRR (and NPV) calculations now commence in the first year with the amounts that were originally in the second year because IRRs are always calculated for future investments and cash flows and thus ignore prior events. That is, the first year starts when the analysis is being done, which is not necessarily the start of the development program. Prior expenses are like water over the dam. Also, the total of the working cash is added into the total cash flow in the last year, whichever it is.

Figures D-5, D-6, and D-7 are the spreadsheets for the three alternatives facing the development team. (Note that all IRRs are higher because the prior year's capital expense is ignored—it's water over the dam.) Thus, these simple calculations reveal that the higher manufacturing cost of alternative X, although it is substantial and must be borne throughout the sales period, is the most attractive alternative. The calculations also spotlight other issues: Is the lower gross margin of alternative X tolerable when judged along with the rest of the manufacturing mix? Is the extra money to fund alternative Z really available? Are people and space to implement this alternative also available? In the case of alternative Y, would a one-year delay in reaching the market truly be acceptable? These factors may outweigh the higher IRRs, which could lead to further exploration to identify other alternatives.

This same kind of analysis can be done to evaluate other managerial choices, such as the trade-off between adding people (for example, more salespersons) or acquiring equipment (for example, cellular telephones for existing salespersons' cars). Another

YEAR =	1	2	3	4	5	6	7	8	9
1 Company Sales		2.00	2.40	2.88	3.46	4.15	4.98	4.98	4.15
2 Manufacturing Cost		1.00	1.20	1.44	1.73	2.07	2.49	2.49	2.07
3 Development Expense	.50								
4 Operating Expense		.40	.48	.58	.69	.83	1.00	1.00	.83
5 Capital Expense									
6 Depreciation	.00	.00	.00	.00	.00	.00	.00	.00	.00
7 Gross Profit	.00	1.00	1.20	1.44	1.73	2.07	2.49	2.49	2.07
8 Before Tax Income	-.50	.60	.72	.86	1.04	1.24	1.49	1.49	1.24
9 Income Tax	-.25	.30	.36	.43	.52	.62	.75	.75	.62
10 Net Income	-.25	.30	.36	.43	.52	.62	.75	.75	.62
11 Operating Cash Flow	-.25	.30	.36	.43	.52	.62	.75	.75	.62
12 Working Cash Required	.00	.60	.12	.14	.17	.21	.25	.00	-.25
13 Total Cash Flow	-.25	-.30	.24	.29	.35	.41	.50	.75	2.12
14 Cumulative Cash Flow	-.25	-.55	-.31	-.02	.32	.74	1.24	1.98	4.10

```
Internal Rate of Return (%) =      52.72


Discounted Cash Flows:
       NPV a 10%          2.04
       NPV a 15%          1.45
       NPV a 20%          1.03
       NPV a 25%           .73
       NPV a 30%           .51
       NPV a 35%           .34
       NPV a 40%           .21
```

FIGURE D-5.   Spreadsheet for alternative X.

example is the choice between designing the initial product to be usable in any country versus the cost of designing a product for, say, the North American market and redesigning it later to meet varied specifications and standards in a myriad of other countries.

PRICING DECISIONS

Pricing is a common problem that may confront you as the manager of new product development. In those situations where price versus volume trade-offs are known (or can be determined by test marketing or conjoint analysis), a DCF provides a convenient format to determine an optimal price point.

EFFECT OF PREMATURE CANCELATION

Very often corporate management will want to understand what happens if the proposed development effort must be terminated early. This is commonly determined by

YEAR =	1	2	3	4	5	6	7	8	9
1 Company Sales			2.00	2.40	2.88	3.46	4.15	4.98	4.98
2 Manufacturing Cost			.80	.96	1.15	1.38	1.66	1.99	1.99
3 Development Expense	.50	.50							
4 Operating Expense			.40	.48	.58	.69	.83	1.00	1.00
5 Capital Expense									
6 Depreciation	.00	.00	.00	.00	.00	.00	.00	.00	.00
7 Gross Profit	.00	.00	1.20	1.44	1.73	2.07	2.49	2.99	2.99
8 Before Tax Income	-.50	-.50	.80	.96	1.15	1.38	1.66	1.99	1.99
9 Income Tax	-.25	-.25	.40	.48	.58	.69	.83	1.00	1.00
10 Net Income	-.25	-.25	.40	.48	.58	.69	.83	1.00	1.00
11 Operating Cash Flow	-.25	-.25	.40	.48	.58	.69	.83	1.00	1.00
12 Working Cash Required	.00	.00	.60	.12	.14	.17	.21	.25	.00
13 Total Cash Flow	-.25	-.25	-.20	.36	.43	.52	.62	.75	2.49
14 Cumulative Cash Flow	-.25	-.50	-.70	-.34	.09	.61	1.23	1.98	4.47

Internal Rate of Return (%) =   46.68

Discounted Cash Flows:

NPV a 10%	2.14
NPV a 15%	1.49
NPV a 20%	1.02
NPV a 25%	.69
NPV a 30%	.45
NPV a 35%	.27
NPV a 40%	.13

FIGURE D-6.   Spreadsheet for alternative Y.

truncating the venture in the ninth or eighth or other year and calculating the IRRs for these shorter efforts. This is another way to judge the risk of a contemplated venture.

HIGHLIGHTS

• IRR or NPV is frequently used to evaluate major investments.

• A spreadsheet program can be used to calculate IRR and NPV and other common financial measures.

• To determine cash flow, which you must do to calculate IRR and NPV, you must determine five estimated quantities: company sales resulting from the program, attendant manufacturing costs, development expense, operating costs, and capital expenditures.

YEAR =	1	2	3	4	5	6	7	8	9
1 Company Sales		2.00	2.40	2.88	3.46	4.15	4.98	4.98	4.15
2 Manufacturing Cost		.80	.96	1.15	1.38	1.66	1.99	1.99	1.66
3 Development Expense	1.20								
4 Operating Expense		.40	.48	.58	.69	.83	1.00	1.00	.83
5 Capital Expense									
6 Depreciation	.00	.00	.00	.00	.00	.00	.00	.00	.00
7 Gross Profit	.00	1.20	1.44	1.73	2.07	2.49	2.99	2.99	2.49
8 Before Tax Income	-1.20	.80	.96	1.15	1.38	1.66	1.99	1.99	1.66
9 Income Tax	-.60	.40	.48	.58	.69	.83	1.00	1.00	.83
10 Net Income	-.60	.40	.48	.58	.69	.83	1.00	1.00	.83
11 Operating Cash Flow	-.60	.40	.48	.58	.69	.83	1.00	1.00	.83
12 Working Cash Required	.00	.60	.12	.14	.17	.21	.25	.00	-.25
13 Total Cash Flow	-.60	-.20	.36	.43	.52	.62	.75	1.00	2.32
14 Cumulative Cash Flow	-.60	-.80	-.44	-.01	.51	1.13	1.88	2.87	5.20

```
Internal Rate of Return (%) =      47.72


Discounted Cash Flows:
    NPV a 10%          2.60
    NPV a 15%          1.84
    NPV a 20%          1.30
    NPV a 25%           .90
    NPV a 30%           .60
    NPV a 35%           .38
    NPV a 40%           .20
```

FIGURE D-7.　　Spreadsheet for alternative Z.

- DCFs (either IRR or NPV) are calculated for future cash flows.
- Financial measures each have a different sensitivity to any change in the five estimated quantities.
- IRR and NPV are the measures usually most sensitive to sales.
- DCF can be used to help with various other new product development management decisions.
- It is crucial to document and understand the underlying assumptions in any financial analysis because the resulting outputs are only significant to the extent that the inputs are realistic.

Notes and References

1. *Business Week,* 19 May 1980, p. 40.
2. *Wall Street Journal,* 30 April 1980.

APPENDIX E

Quality Function Deployment— The House of Quality

QFD is a graphic planning and documentation system for deploying the voice of the customer. Its core value lies in its holistic approach that facilitates communication between different functions engaged in the new product development process.

QFD is a graphical planning and documentation system that greatly facilitates the team effort. Simply laying market requirements alongside design elements is a powerful way to bring the multifunctional team together to make the necessary product decisions and trade-offs.

Figure 4-3 shows an overview of the QFD "house of quality." We investigate each "room" of this house as we work our way through a sample design project. Figure E-1 shows the starting point of our exercise—a blank matrix illustrating WHATs and HOWs.* The matrix separates WHAT the customer wants from HOW it will be accomplished. Very often these two design considerations can get confused. QFD does an excellent job of keeping them separate and thereby facilitating the design and product development processes.

We take a step-by-step journey through the house of quality by using QFD to help in the design of a fictitious laser printer. At the completion of this exercise, you should appreciate QFD and how it works. Remember, there is no magic here. The value added of QFD is its structure, graphical representation, and documentation capabilities.

Figure E-2 shows the next step in the design of our laser printer. We have developed a

---

*These QFD matrices were produced using a program designed for the QFD process, "QFD Designer." It is available from Qualisoft Corporation, 7395 Bridgeway West, West Bloomfield, MI 48322. Another popular QFD software program, "QFD/Capture," is produced by International TechneGroup Incorporated, 5303 DuPont Circle, Milford, OH 45150.

QFD Example
New Laser Printer

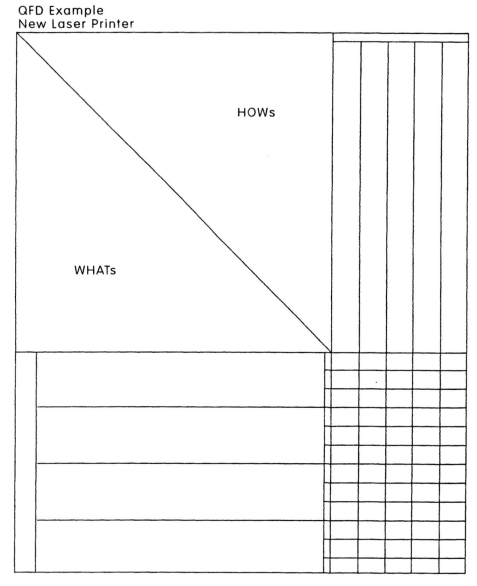

FIGURE E-1.   House of Quality: blank matrix.

list of WHATs, including their importance from a market research project with existing
and prospective customers. These four WHATs are a compilation of customer responses
for the attributes they consider when purchasing a laser printer. These responses could
have been obtained from a number of market research approaches similar to those
discussed in Chapter 4. It is important to do a good job here–no marketing or engineering guesses of what the customer wants. Take the time and effort to go to the customer
because these WHATs will drive the rest of the QFD process.

QFD Example
New Laser Printer

WHATs \ HOWs	IMPORTANCE							
Compatible	4							
Print Quality	5							
Ease of Use	3							
Productivity	5							

FIGURE E-2. House of Quality: new laser printer, WHATs.

In team meetings, we have developed the basic design elements or HOWs (Figure E-3) we plan to implement in order to meet these customer needs. These basic HOWs meet three necessary criteria:

1. They should be measurable and controllable.
2. No parts or process are allowed (at this stage).
3. They should be benchmarkable (measured against competing firms or products).

Generally, it is useful to brainstorm several HOWs that address each customer requirement (WHAT). Unless closely controlled, however, the complexity of the matrix can easily get out of hand and become unwieldy. If the matrix does become too large, it may be desirable to break out groups of related HOWs into subsystems for additional analysis.

The value added for this basic "house" is the matrix intersection of the WHATs and the HOWs. The multifunction team has evaluated each intersection with a two-step process:

1. Determine if this HOW has any effect on the WHAT.
2. Determine the level of effect. There are three possible levels: weak, medium, and strong.

QFD Example
New Laser Printer

DIRECTION OF IMPROVEMENT		○	↑	↑	○	↓	↑	↑
HOWs / WHATs	IMPORTANCE	Postscript Compatible	Resolution	Edge Sharpness	Duplex Printing	Hrs. Training Req'd.	Speed Text	Speed Graphics
Compatible	4							
Print Quality	5							
Ease of Use	3							
Productivity	5							

FIGURE E-3.
House of Quality: new laser printer, HOWs.

QFD Example
New Laser Printer

DIRECTION OF IMPROVEMENT		○	↑	↑	○	↓	↑	↑
HOWs / WHATs	IMPORTANCE	Postscript Compatible	Resolution	Edge Sharpness	Duplex Printing	Hrs. Training Req'd.	Speed Text	Speed Graphics
Compatible	4	●						
Print Quality	5	△	●	●				
Ease of Use	3				○	●		
Productivity	5				●	△	●	●

MATRIX		WEIGHTS	ARROWS	
Strong	●	9	Maximize	↑
Medium	○	3	Minimize	↓
Weak	△	1	Nominal	○

FIGURE E-4.
House of Quality: new laser printer, relationship between WHATs and HOWs.

As shown in Figure E-4, we have captured the results of this evaluation. We have also agreed on the direction of improvement for each of the design considerations. This is an important step in the QFD process because the relationships between the WHATs and HOWs will be an important guide for the product design. For example, the availability of Postscript™ has both a strong impact on compatibility and a weak impact on print quality. Both these considerations will be used to figure the overall contribution of the Postscript™ page description language to meeting customer requirements. Important HOWs will become WHATs in subsequent stages as we continue to deploy the voice of the customer to more detailed design and production considerations.

In Figure E-5, we have expanded the basic QFD house for our laser printer to a much more detailed evaluation and documentation effort. The additional data are in rooms added to our basic house. Keep in mind that these rooms are not a standard form to be completed. Rather, they should be viewed as a convenient way to document and consolidate the market and design requirements for the new product. Some of the important rooms added to this expanded house include the following:

Targets. These values represent the "how much" associated with each HOW. Target values should be looked at as the value necessary to satisfy the customer. The values should represent the "benchmark" of what will be required to meet customer requirements when the new product is available. They are best established by the designers with the help of the team. Rather than get bogged down in the "right" target, it is frequently helpful to establish preliminary values to be reviewed later in the analysis.

The roof. The roof of the house of quality provides for the trade-off and evaluation of conflicts that arise between the HOWs. For example, a strong negative relationship is indicated between postscript compatibility and graphics speed of the printer. A strong positive (reinforcing) relationship was determined to exist between postscript compatibility and resolution. These relationships can be helpful to the team when considering design trade-offs.

Figure E-5 also shows the calculation of the absolute and relative importance of each design element (HOW) in addressing customer requirements (WHAT). This calculation is simply a summation of the matrix weights multiplied by the customer importance rating. Duplex Printing, for example, got a score of 54 (Productivity Importance, 5 × Matrix Weight, 9 = 45, plus Ease of Use Importance, 3 × Matrix Weight, 3 = 9). This results in a Relative Importance of 17 percent higher than any other HOW.

Note that the team also estimated the degree of organizational difficulty associated with the accomplishment of each HOW. Duplex Printing and Postscript Compatibility both received a rating of 5, indicating they will be the most difficult design elements to achieve.

QFD Example
New Laser Printer

DIRECTION OF IMPROVEMENT		○	▲	▲	○	▼	▲	▲
HOWs / WHATs	IMPORTANCE	Postscript Compatible	Resolution	Edge Sharpness	Duplex Printing	Hrs. Training Req'd.	Speed Text	Speed Graphics
Compatible	4	◉						
Print Quality	5	△	◉	◉				
Ease of Use	3			.	○	◉		
Productivity	5				◉	△	◉	◉
ORGANIZATIONAL DIFFICULTY		5	3	3	5	3	3	3
TARGETS		YES	400 dpi.	.01 mm variation	Automatic Duplex	16 hrs. maximum	10 ppm	5 ppm
ABSOLUTE IMPORTANCE		41	45	45	54	32	45	45
RELATIVE IMPORTANCE		13%	14%	14%	17%	10%	14%	14%

ROOF	MATRIX	WEIGHTS	ARROWS
Strong Pos. ◉	Strong ◉	9	Maximize ▲
Positive ○	Medium ○	3	Minimize ▼
Negative ✕	Weak △	1	Nominal ○
Strong Neg. ※			

FIGURE E-5.   House of Quality: new laser printer, roof and design targets.

QFD Example
New Laser Printer

DIRECTION OF IMPROVEMENT	IMPORTANCE	Postscript Compatible	Resolution	Edge Sharpness	Duplex Printing	Hrs. Training Req'd.	Speed Text	Speed Graphics	CUSTOMER RATING — Our new Product ▲ / Product A □ / Product B ○ (0 1 2 3 4 5)
HOWs / WHATs		○	▲	▲	○	▼	▲	▲	
Compatible	4	◉							▲ (0) ... □
Print Quality	5	△	◉	◉					□ ▲ ○
Ease of Use	3				○	◉			▲ □○
Productivity	5				◉	△	◉	◉	▲ ○ □
ORGANIZATIONAL DIFFICULTY		5	3	3	5	3	3	3	
TARGETS		YES	400 dpi.	.01 mm variation	Automatic Duplex	16 hrs. maximum	10 ppm	5 ppm	

ENGINEERING COMPETITIVE ASSESSMENT

▲ Our new Prod	5
□ Product A	4
○ Product B	3
	2
	1
	0

	Postscript Compatible	Resolution	Edge Sharpness	Duplex Printing	Hrs. Training Req'd.	Speed Text	Speed Graphics
ABSOLUTE IMPORTANCE	41	45	45	54	32	45	45
RELATIVE IMPORTANCE	13%	14%	14%	17%	10%	14%	14%

ROOF		MATRIX		WEIGHTS	ARROWS	
Strong Pos.	◉	Strong	◉	9	Maximize	▲
Positive	○	Medium	○	3	Minimize	▼
Negative	×	Weak	△	1	Nominal	○
Strong Neg.	※					

FIGURE E-6.   House of Quality: new laser printer, customer and competitive ratings.

We were fortunate to have recently completed both an engineering and customer assessment of our design and competitive products A and B. Note that the engineering assessment measured the HOWs, whereas the customer assessment responded to the WHATs for the products tested.

As shown in Figure E-6, the results of these two studies are graphically represented in rooms added to our house: the customer rating room and the engineering competitive assessment room. Our new laser printer product fared better in engineering competitive assessment than in customer ratings. The QFD matrix can be a big help in finding out why these apparent inconsistencies exist. For example, although our engineering test shows the new product to be a clear winner over competitive products in both text and graphic speed, customers rated the product last in productivity. It is important to understand the reasons behind this discrepancy. Perhaps some other design characteristics of our product severely compromise speed in actual practice. Perhaps the customer's benchmarks are very different from ours. It is important to find out why these results differ.

QFD PHASES

The deployment aspect of QFD consists of driving the customer's critical HOWs to subsequent phases of the product design and manufacturing processes. This process is shown in Figures E-7, E-8, and E-9. Note that important HOWs become the WHATs of the subsequent phase as we address increasingly detailed elements of the product development and manufacturing processes. For our laser printer, we have driven a customer requirement for productivity in their laser printer:

- To design of automatic duplex printing
- To critical parts to accomplish that design
- To the important design elements of those parts
- To control of the manufacturing process used to produce the parts

The QFD must be directed at the *important* (as determined by the customer) product design elements. The value of the QFD process is in its ability to prioritize and focus the multifunctional team process in bringing higher value to the customer.

PHASE 2

As shown in Figure E-6, duplex printing—the ability to print automatically on both sides of a piece of paper—is a design approach that has both a high relative importance and organizational difficulty. This qualifies duplex printing as a design element requiring additional analysis by the team.

In Figure E-7, parts deployment, we have evaluated the design for automatic duplexing and determined there are three critical parts involved:

1. The rollers that keep the paper moving through the printer
2. The turners that reverse the paper path for temporary storage while the second half of the page is being prepared for printing
3. The paper-holding tray

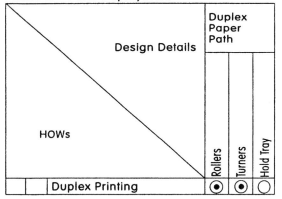

QFD Example
New Laser Printer
Phase 2 – Parts Deployment

FIGURE E-7.   House of Quality: new laser printer, phase 2—parts deployment.

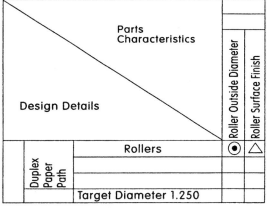

QFD Example
New Laser Printer
Phase 3 – Process Planning

FIGURE E-8.   House of Quality: new laser printer, phase 3—process planning.

QFD Example
New Laser Printer
Phase 4 – Production Planning

DIRECTION OF IMPROVEMENT				Lathe Knob Control	Operator Training	Lathe Maintenance
Production Process						
Parts Characteristics						
Roller Outside Diameter				⊙	○	○

MATRIX		WEIGHTS	ARROWS	
Strong	⊙	9	Maximize	▲
Medium	○	3	Minimize	▼
Weak	△	1	Nominal	○

FIGURE E-9.   House of Quality: new laser printer, phase 4—production planning.

We have further identified the rollers as the critical design element requiring additional investigation.

PHASE 3

In Figure E-8, we have determined that the outside diameter is a key roller design element in achieving smooth, no jam, duplex printing operation. We looked at two roller design elements: outside diameter and surface finish. The team decided the roller outside diameter was more critical, requiring close tolerances.

PHASE 4

Figure E-9 shows the result of the team's assessment of the manufacturing process to produce rollers consistently within tolerance. Investigation by the manufacturing team member indicates that the lathe used to produce the rollers and the lathe knob control mechanism to cut the proper diameter are the critical elements in the production process.

We have deployed the voice of the customer through the design to the manufacturing process. We were guided along the way by focusing on only critical, high-priority elements that contributed to improving our response to customer requirements. At this point, the team may deem it appropriate to cycle back to a previous stage and pick the next priority item. The QFD process becomes an important part of the continuous improvement philosophy we have discussed in other parts of this book.

OTHER APPLICATIONS OF QFD

Although QFD has been primarily used to address the design and manufacture of hardware products, its capabilities go well beyond that single application. Figures E-10 and E-11 illustrate how QFD was used to design a marketing project. This application uses QFD to help design a market research project to evaluate how the market for multimedia applications will influence system design. Outlining the steps in the QFD graphic format reduces a relatively complex project to simpler, easier to understand elements.

Figure E-10 shows how market segments (in this case, functional department types) might relate in their use of multimedia for some common applications. Figure E-10 further defines and documents several important aspects of the project, including the following:

We've defined "market segment" as functional departments. This is an important step that will have significant impact on the acquisition and analysis of market data.
We've defined multimedia applications in five areas. We further provided specific examples under each application category heading.

QFD Example
Market Research Project
Phase 1

MARKET SEGMENT / MULTIMEDIA APPLICATION	IMPORTANCE	Education & Training		Help		Business Communication					Storage & Retrieval			Direct Production	
		Employee Training	Customer Training	Sales Kiosks	Customer Service	Presentations	Correspondence	Messages	Documents	Meetings	Employee Records	Medical Records	Engineering Documentation	Advertising	Storyboards
Manufacturing & Operations		○					○		○	○	△		◉		
							○								
Marketing & Sales		○	◉	◉		◉	◉		◉	◉				◉	○
Human Resources & Training		◉							◉		◉	◉			
R & D and Engineering		○							○	◉			◉		
Customer Service & Support		○	○		◉		○	△	○	○			○		
Finance & Administration		○					◉	◉	◉	○	○				

MATRIX		WEIGHTS
Strong	◉	9
Medium	○	3
Weak	△	1

FIGURE E-10.   House of Quality: multimedia research project, market segment vs. application.

The team has documented which market segments are expected to have a strong, medium, or weak interest in specific multimedia applications.

Although this structure remains to be verified by market research, the value of establishing and documenting this common set of assumptions can go a long way toward reducing confusion that can exist in the design and basic assumptions behind a large market research project.

Similar to our laser printer example, Figure E-11 refines the process by relating two important HOWs (in this case, business communication and storage and retrieval applications) to the specific capabilities of multimedia systems. We can continue to deploy the important HOWs to additional levels of refinement and definition. In Figure E-11, we have accomplished the following:

QFD Example
Market Research Project
Phase 2

Multimedia System Capabilities / MULTIMEDIA APPLICATION	Graphic Capabilities				Audio Capabilities			Video Capabilities		
	Text & Data	Simple Graphics	Complex Graphics	Animation	Basic Audio	Simulated Voice	Full Stereo	Freeze Frame	Slow Motion	Full Motion
Presentations	◉	◉	△	○		○	○		○	◉
Correspondence	◉	○	△			○				
Messages	◉					◉				
Documents	◉	◉				○				
Meetings	○				◉	△			○	◉
Employee Records	◉					◉				◉
Medical REcords	◉		◉							
Engineering Documentation			◉							

(Business Communication: Presentations, Correspondence, Messages, Documents, Meetings; Storage & Retrival: Employee Records, Medical REcords, Engineering Documentation)

MATRIX		WEIGHTS
Strong	◉	9
Medium	○	3
Weak	△	1

FIGURE E-11.   House of Quality: multimedia research project, application vs. system capabilities.

We have decided that storage and retrieval and business communication applications
are the ones we want to investigate further. We have made a conscious, documented
decision to exclude the other applications either because they are not as attractive
to customers or on some other basis.

We have defined in broad terms graphic, audio, and video elements that we will use to
define a "multimedia system."

We have evaluated and documented how strongly the specific system capabilities will
relate to the storage and retrieval and business communication applications we
have under consideration.

At this point, the team has constructed a guideline to be used in the acquisition and
analysis of market information. The refinement of the process could easily continue to
address a specific design technology or product area such as storage media, software,
work station capability, and so forth.

HIGHLIGHTS

- The flexibility of QFD should not be confined to the design and manufacturing of hardware products. The powerful graphic–matrix elements of the process can be used to understand and document many of the other complex analyses and decisions that will arise in the new product development process.

In a recent study of QFD across 35 new product development teams the following suggestions were offered to increase the successful use of QFD by organizations:

- Treat QFD as an investment not an expense.

- Apply QFD to less complex projects that involve incremental change to an existing design (at least initially until you are familiar with its capabilities and limitations).

- The new product development *team* should provide the incentive for the use of QFD not senior management.

- View QFD as a tool in the new product development process not as an end objective.

- QFD does not provide better market data or improved design; however, by aiming at the process and the *use* of market data in design, QFD ends up improving both.

Project Management Software

There is a wide variety of microcomputer-based (and other) project management software that can be used advantageously. Its primary value is for planning and explaining the new product development schedule so it is consistent with the multifunctional resources that will be utilized.

There is an old saying, "Failure to plan is planning to fail." That pretty much sums up the use of project management software on a new product development project. It's not that you cannot be successful with a "seat of the pants" schedule plan, but success is less certain. Conversely, using project management software will not assure that the new product will be developed on time or be successful in the competitive marketplace. But this class of software is widely available, some of it is reasonably convenient to use and provides helpful output, and it is therefore something with which the new product development project manager should be acquainted. Unless you have a simpler method to determine the essential information such software can furnish and provide it to the multifunctional project team, we urge you to use it on your new product development project.

Given that there are dozens of available packages, each of which provides an upgrade or revision frequently, it is impossible to write anything definitive that will remain enduringly current. Thus, we merely want to provide a few specific examples of output from representative microcomputer-based project management software. The following are the general inputs and outputs for this kind of software:

Inputs for each task or event
  Name or identification
  WBS detail (if used)
  Duration (events normally have zero time duration)

Predecessor and successor tasks and events
Resources and their rate of use

Outputs for entire project
WBS chart
Gantt or bar chart
Network diagram of some sort (PERT, precedence, or TBAOA)
Resource usage by time period
Cost projection

Figure F-1 is a Gantt or bar chart that shows milestones and lists WBS identifiers. Figure F-2 is the WBS for the example new product development project we used in Chapters 6 and 7, and Figure F-3 is the WBS display for a different (but similar) project. Most packages also provide a network precedence diagram (called, usually mistakenly, a PERT diagram), such as Figure F-4, that shows predecessor and successor tasks without a linear time scale.

We are strong advocates of what we have called (in Chapter 7) time-based activity-on arrow (TBAOA) network diagram schedules. Some, but not all, microcomputer project management software can provide this. Unfortunately, there is no standard language, and some software packages—including several very popular ones—have features (for example, "time-based PERT," Figure F-5) that sound like TBAOA but are not.

Figures F-6, F-7, and F-8 are examples of microcoputer-based project management software that provides TBAOA today. These three software packages are not the only ones that provide TBAOA, and our choice was made primarily for convenience. However, the three examples are reasonably representative of mass market (and other) software. Figures F-6 and F-7 are TBAOA diagrams for the example new product development project introduced in Figure 7-2 and Table 7-2. TBAOA may also be thought of as Gantt or bar charts in which each bar is linked to its predecessors and successors.

Some software (ViewPoint, for example, Figure F-8) allows you to place the bars at levels of your choosing, so you can have a band or zone for all the work by one department, another band for another department, and so on. You can use this same technique to display the involvement of key contractors (or your customer, if the new product development effort is for an OEM). Using such bands can clarify who is to do what when, which is a great aid for coordination. Conversely, this format can add extra vertical connections to the display, and that clutter may become confusing.

Figure F-9 is a display showing resource availability (positive numbers at the bottom of the figure) and deficits (negative numbers), based on the resource requirements shown in Table 8-1 for the example new product development project of figure 7-2 and Table 7-2. Whenever there are resource deficits (mechanical and electrical engineers, in the example), the new product development schedule is unrealistic. Thus, this is a danger signal for the new product development project manager and the entire multi-functional project team. Unless more resources are obtained or the plan is changed, the development schedule will not be met.

There are many other options for software output, which we have not illustrated here. We believe the essential ones for the new product development project manager are TBAOA and resource use (or deficits).

SOME MICROCOMPUTER-BASED PROJECT MANAGEMENT SOFTWARE PROVIDING TBAOA

The following five vendors sell microcomputer project management software that can create time-based activity-on-arrow critical path networks. Other vendors also provide software products with this capability. Citation in this list is neither an endorsement nor a recommendation; the list—current in late 1992—is provided solely as a convenience.

Project Workbench 3.1 with Project Graphics, Applied Business Technology, 361 Broadway, New York, New York 10013, (800) 477-6532 or (212) 219-8945

SuperProject 2.0, Computer Associates, 2195 Fortune Drive, San Jose, California 95131, (408) 432-1727

SureTrak Project Scheduler, Primavera Systems, 1574 West South Street, Salt Lake City, Utah 84104, (810) 973-9725

Texim Project, Welcom Software Technology, 15995 North Barkers Landing, Suite 275, Houston, Texas 77079, (713) 558-0514

ViewPoint, Computer Aided Management, Suite 210, 1318 Redwood Way, Petaluma, California 94954-9935, (707) 795-4100

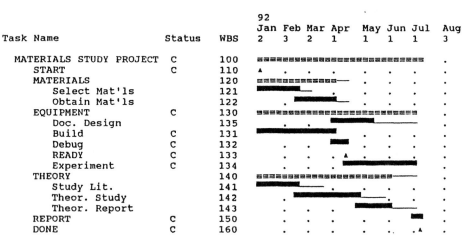

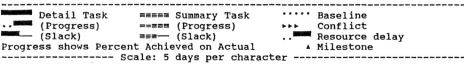

```
Schedule Name :
Responsible   :
As-of Date    : 6-Aug-90  9:00am       Schedule File : MSPROJ


                                        92
                                        Jan Feb Mar Apr  May Jun Jul  Aug
Task Name                Status  WBS    2   3   2   1    1   1   1    3

    MATERIALS STUDY PROJECT   C   100
        START                 C   110
        MATERIALS                 120
            Select Mat'ls         121
            Obtain Mat'ls         122
        EQUIPMENT             C   130
            Doc. Design           135
            Build             C   131
            Debug             C   132
            READY             C   133
            Experiment        C   134
        THEORY                    140
            Study Lit.            141
            Theor. Study          142
            Theor. Report         143
        REPORT                C   150
        DONE                  C   160
```

TIME LINE Gantt Chart Report, Strip 1

FIGURE F-1. A bar chart with milestones and work breakdown structure task identifiers using Time Line software.

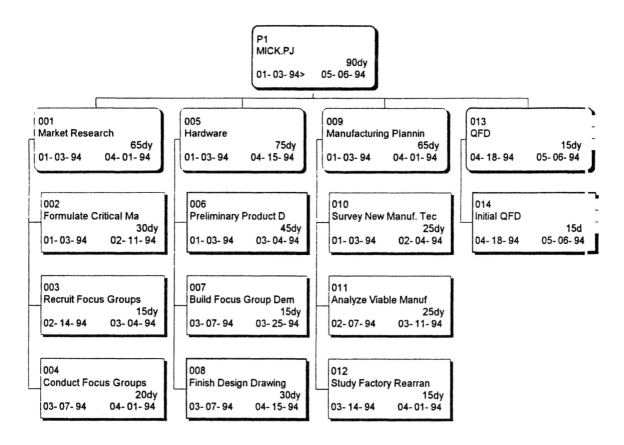

FIGURE F-2.   A work breakdown structure for the example project of Figure 7-2 and Table 7-2, using SuperProject 2.0 software.

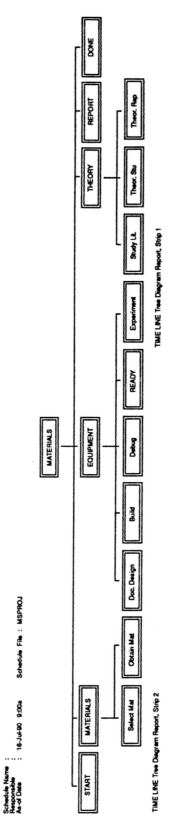

FIGURE F-3.   A work breakdown structure, as portrayed in Time Line software.

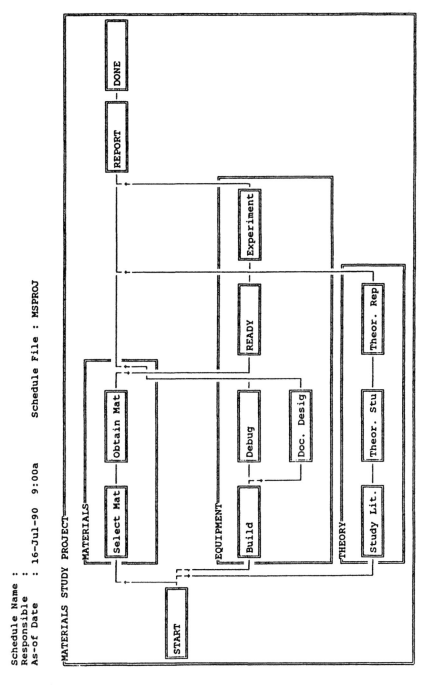

FIGURE F-4.   A so-called PERT chart, which is really a hybrid network diagram with both events and activities in the nodes (boxes), prepared with Time Line software.

244

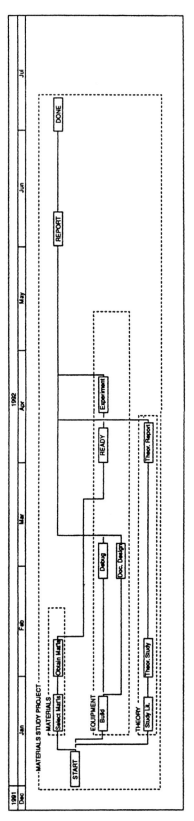

FIGURE F-5.   A "time-based PERT" chart. Note that the activities and events are not linearly scaled in time.

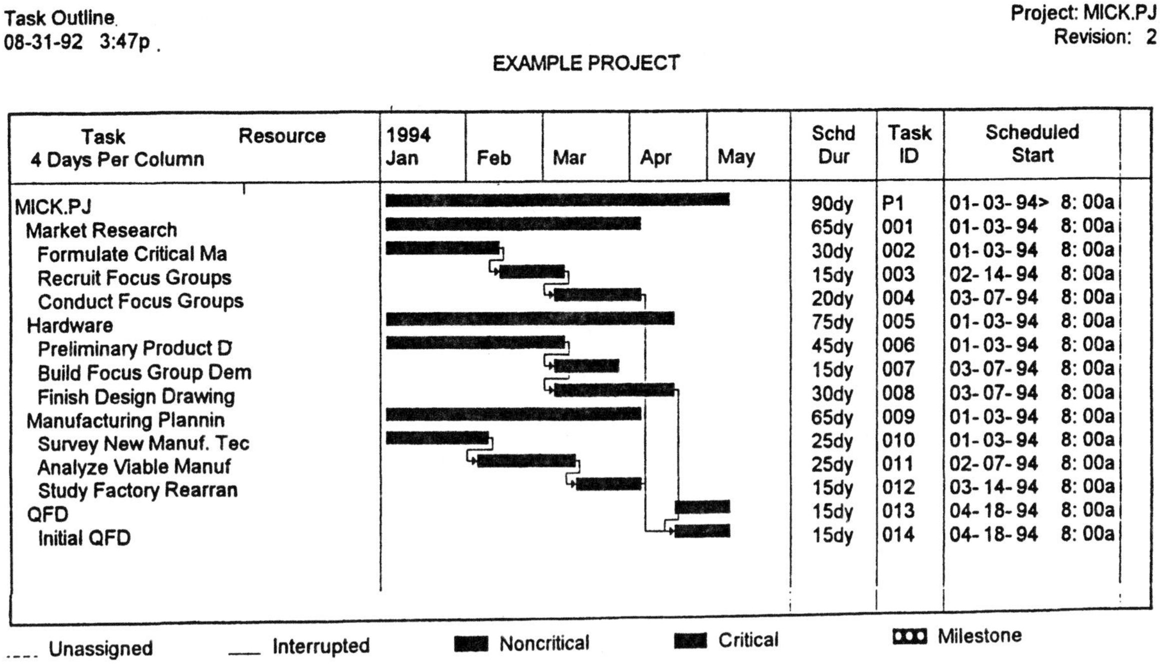
Task Outline
08-31-92 3:47p
Project: MICK.PJ
Revision: 2
EXAMPLE PROJECT
Task
4 Days Per Column
Resource
1994
Jan
Feb
Mar
Apr
May
Schd Dur
Task ID
Scheduled Start
MICK.PJ 90dy P1 01-03-94> 8:00a
Market Research 65dy 001 01-03-94 8:00a
Formulate Critical Ma 30dy 002 01-03-94 8:00a
Recruit Focus Groups 15dy 003 02-14-94 8:00a
Conduct Focus Groups 20dy 004 03-07-94 8:00a
Hardware 75dy 005 01-03-94 8:00a
Preliminary Product D 45dy 006 01-03-94 8:00a
Build Focus Group Dem 15dy 007 03-07-94 8:00a
Finish Design Drawing 30dy 008 03-07-94 8:00a
Manufacturing Plannin 65dy 009 01-03-94 8:00a
Survey New Manuf. Tec 25dy 010 01-03-94 8:00a
Analyze Viable Manuf 25dy 011 02-07-94 8:00a
Study Factory Rearran 15dy 012 03-14-94 8:00a
QFD 15dy 013 04-18-94 8:00a
Initial QFD 15dy 014 04-18-94 8:00a
Unassigned
Interrupted
Noncritical
Critical
Milestone

FIGURE F-6. A time-based activity-on-arrow schedule for the example project of Figure 7-2 and Table 7-2, as portrayed in SuperProject 2.0 software.

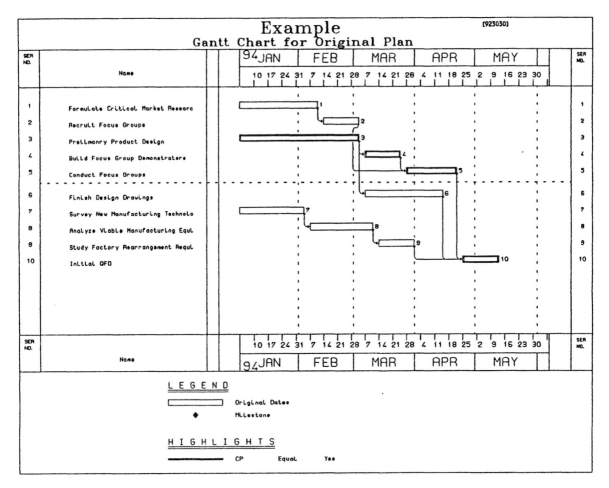

FIGURE F-7.   A time-based activity-on-arrow schedule for the example project of Figure 7-2 and Table 7-2, as portrayed in Project Workbench 3.1 with Project Graphics.

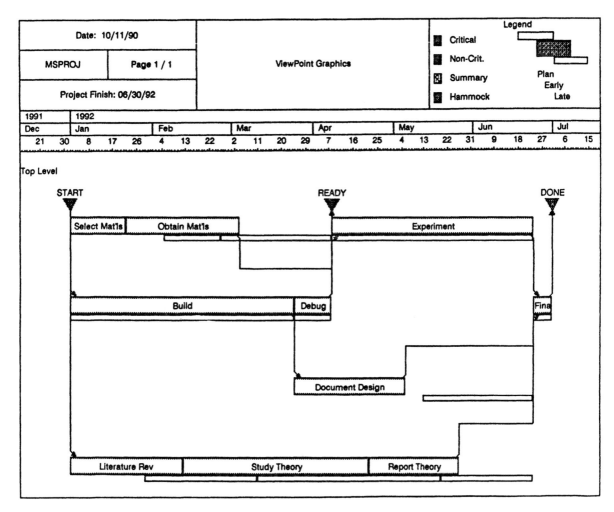

FIGURE F-8.   A time-based activity-on-arrow schedule prepared with ViewPoint. (Note that this output display is called a "time-based precedence diagram" in that software.)

12-24-92	Gantt Chart	Page 1

Task Name	Usage	Res List	January 1994 · · · February 1994 · · · March 1994 · · · April 1994 · · · May 1994
Formulate Critical Market	480	MKT1 MKT2	
Recruit Focus Groups	120	MKT1	
Preliminary Product Design	1,440	ME1 ME2 EE	
Build Focus Group Demonstration	960	ME1 ME2 T1	
Conduct Focus Groups	640	MKT1 MKT2	
Finish Design Drawings	480	ME1 EE1	
Survey New Manufacturing	200	MFE	
Analyze Viable Manufacturing	200	MFE	
Study Factory Rearrangement	120	MFE	
Initial QFD	840	MKT1 MKT2	

12-24-92	Resource Spreadsheet	Page 1

Resource Summary / Unused Availability	Avail (hrs/d)	Res Abbv	J10	J17	J24	J31	F7	F14	F21	F28	M7	M14	M21	M28	A4	A11	A18	A25	Y2	Y9	Y16	Y23
Marketing	8.0	MKT1									40	40	40	-40	-40	-40	-40				40	40
Marketing	8.0	MKT2						40	40	40	40	40	40	-40	-40	-40	-40				40	40
Mechanical Engineer	8.0	ME1									-40	-40	-40					40			40	40
Mechanical Engineer	8.0	ME2												40	40	40	40				40	40
Electrical Engineer	8.0	EE1									-40	-40	-40					40			40	40
Electrical Engineer	8.0	EE2													40	40	40				40	40
Technician	8.0	T1	40	40	40	40	40	40	40	40	40				40	40	40	40	40	40	40	40
Technician	8.0	T2	40	40	40	40	40	40	40	40	40				40	40	40	40	40	40	40	40
Technician	8.0	T3	40	40	40	40	40	40	40	40	40				40	40	40	40	40	40	40	40
Technician	8.0	T4	40	40	40	40	40	40	40	40	40				40	40	40	40	40	40	40	40
Manufacturing Engineer	8.0	MFE													40	40	40				40	40

FIGURE F-9.   Resource availability and deficits for the example project of Figure 7-2 and Table 7-2, based on the resource requirements of Table 8-1, as portrayed in Project Workbench 3.1.

Further Reading

PART 1

M. D. Rosenau. *Faster New Product Development: Getting the Right Product to Market Quickly.* New York: AMACOM, 1990, pp. 3-98.

M. D. Rosenau. *Successful Project Management,* 2nd ed. New York: Van Nostrand Reinhold, 1992, pp. 1-11.

S. R. Rosenthal. *Effective Product Design and Development.* Homewood, IL: Business One Irwin, 1992, pp. 1-9.

P. G. Smith and D. G. Reinersten. *Developing Products in Half the Time.* New York: Van Nostrand Reinhold, 1991, pp. 1-15, 61-79.

S. C. Wheelwright and K. B. Clark. *Revolutionizing Product Development: Quantum Leaps in Speed, Efficiency, and Quality.* New York: Free Press (Macmillan), 1992, pp. 1-56, and 133-164.

PART 2

Books

R. D. Buzzell and B. T. Gale. *The PIMS Principles.* New York: Free Press, 1987.

R. G. Cooper. *Winning at New Products.* Reading, MA: Addison-Wesley, 1986, pp. 48-65.

M. D. Rosenau. *Faster New Product Development: Getting the Right Product to Market Quickly.* New York: AMACOM, 1990, pp. 3-67.

M. D. Rosenau. *Successful Project Management,* 2nd ed. New York: Van Nostrand Reinhold, 1992, pp. 13-38.

S. R. Rosenthal. *Effective Product Design and Development.* Homewood, IL: Business One Irwin, 1992, pp. 17-49.

S. C. Wheelwright and K. B. Clark. *Revolutionizing Product Development: Quantum Leaps in Speed, Efficiency, and Quality.* New York: Free Press (Macmillan), 1992, pp. 57-132.

Articles

J. R. Hauser and D. Clausing. "The House of Quality." *Harvard Business Review,* May–June 1988, pp. 63–73.

This is the best layperson's overview of QFD.

J. Holusha. "Raising Quality: Consumers Star." *New York Times,* 5 January 1989, p. C1ff.

This brief news article describes the growing use of QFD and its benefits.

P. H. Lewis. "Software to Help Introduce Products." *New York Times,* 6 October 1991, sec. 3.

Business Insights™ is an expert system software package based on the experience of thirty business strategy and product planning experts. The product provides an exhaustive questionnaire to profile either a product- or service-based business. It can be very helpful as a completeness check and to test the level of understanding behind key factors and assumptions. It also has provisions to use the data directly in developing a business plan for the new product or service.

J. F. Nunamaker. "Workgroup Computing." *Corporate Computing,* August 1992, pp. 196–198.

This article describes the new range of software tools to help facilitate the multifunctional team decision process.

S. Schoeffler, R. D. Buzzell, and D. F. Heany. "Impact of Strategic Planning on Profit Performance." *Harvard Business Review,* March–April 1974, pp. 137–145.

PIMS (Profit Impact of Marketing Strategy) was started by the General Electric Company in 1960 to find better ways to predict profitability. It consists of a collection of hundreds of data items on hundreds of businesses. Because of the size and extent of these data bases, PIMS can provide a reference point on the business experience of companies with products closely matched to most users. It addresses operational issues with an emphasis on new products. The core of the analysis measures the impact on profitability of a number of new product strategies, including market share, rate of new product introduction, breadth of product line, and relative product quality.

G. Van Treeck and R. Thackeray. "Quality Function Deployment at Digital Equipment Corp." *Concurrent Engineering,* January–February 1991, pp. 14–20.

This short article explains the use of QFD for software products.

G. Watson. "Hoshin Kanri—Japan's System for Successful Long-Term Planning." *Boardroom Reports,* 15 April 1992, pp. 3–4.

This is a brief description of Japan's disciplined approach to continuous improvement of the strategic planning process. Hoshin Planning is a disciplined approach to planning that ties together long-term goals and daily operations. Hoshin Planning or "Hoshin Kanri" uses forms to achieve continuous improvement by focusing on variations between the plan and actual results.

PART 3

M. D. Rosenau. *Faster New Product Development: Getting the Right Product to Market Quickly.* New York: AMACOM, 1990, pp. 135–157, 258–263.

M. D. Rosenau. *Successful Project Management,* 2nd ed. New York: Van Nostrand Reinhold, 1992, pp. 49–150.

S. R. Rosenthal. *Effective Product Design and Development.* Homewood, IL: Business One Irwin, 1992, pp. 177–122.

P. G. Smith and D. G. Reinertsen. *Developing Products in Half the Time.* New York: Van Nostrand Reinhold, 1991, pp. 189–221.

PART 4

Books

M. D. Rosenau. *Faster New Product Development: Getting the Right Product to Market Quickly.* New York: AMACOM, 1990, pp. 101–121.

M. D. Rosenau. *Successful Project Management,* 2nd ed. New York: Van Nostrand Reinhold, 1992, pp. 153–193.

S. R. Rosenthal. *Effective Product Design and Development.* Homewood, IL: Business One Irwin, 1992, pp. 85–116.

P. G. Smith and D. G. Reinertsen. *Developing Products in Half the Time.* New York: Van Nostrand Reinhold, 1991, pp. 111–151.

S. C. Wheelwright and K. B. Clark. *Revolutionizing Product Development: Quantum Leaps in Speed, Efficiency, and Quality.* New York: Free Press (Macmillan), 1992, pp. 165–217.

Articles

M. Eames and D. Wilemon. "Determinants of Cross-Functional Cooperation in Technology-Based Organizations." *Journal of Engineering and Technology Management,* Vol. 7, 1991, pp. 229–250.

Based on a field study, this article discusses those issues that impede cross-functional integration and explains what managers can do to improve cooperation.

C. G. King. "Multi-Discipline Teams: A Fundamental Element of the Program Management Function." *PM NetWork,* August 1992, pp. 13–22.

The article provides a thorough description of how Boeing forms, trains, uses, and empowers multifunctional teams to work both independently and in linked arrangements to handle its commercial airliner and varied military aerospace business.

H. J. Thamhain and D. L. Wilemon. "Conflict Management in Project Life Cycles." *Sloan Management Review,* Vol. 16, No. 3, Spring 1975, pp. 31–50.

This article reviews the kinds of conflict project managers encounter and some ways to cope with it.

H. J. Thamhain and D. L. Wilemon. "Leadership, Conflict and Program Management Effectiveness." *Sloan Management Review,* Vol. 19, No. 1, Fall 1977, pp. 69–89.

This article has research data on effective management techniques for project managers.

PART 5

M. D. Rosenau. *Faster New Product Development: Getting the Right Product to Market Quickly.* New York: AMACOM, 1990, pp. 135–157, 164–180.

M. D. Rosenau. *Successful Project Management,* 2nd ed. New York: Van Nostrand Reinhold, 1992, pp. 197–250.

S. R. Rosenthal. *Effective Product Design and Development.* Homewood, IL: Business One Irwin, 1992, pp. 32–35, 73–83.

P. G. Smith and D. G. Reinertsen. *Developing Products in Half the Time.* New York: Van Nostrand Reinhold, 1991, pp. 169–186.

S. C. Wheelwright and K. B. Clark. *Revolutionizing Product Development: Quantum Leaps in Speed, Efficiency, and Quality.* New York: Free Press (Macmillan), 1992, pp. 218–244.

PART 6

M. D. Rosenau. *Faster New Product Development: Getting the Right Product to Market Quickly.* New York: AMACOM, 1990, pp. 181–186.

M. D. Rosenau. *Successful Project Management,* 2nd ed. New York: Van Nostrand Reinhold, 1992, pp. 253–263.

S. R. Rosenthal. *Effective Product Design and Development.* Homewood, IL: Business One Irwin, 1992, pp. 284–289, 291–301.

S. C. Wheelwright and K. B. Clark. *Revolutionizing Product Development: Quantum Leaps in Speed, Efficiency, and Quality.* New York: Free Press (Macmillan), 1992, pp. 284–337.

APPENDIX D

Book

P. G. Smith and D. G. Reinertsen. *Developing Products in Half the Time.* New York: Van Nostrand Reinhold, 1991, pp. 17–41.

Article

C. Y. Baldwin. "How Capital Budgeting Deters Innovation—And What To Do About It." *Research-Technology Management,* November–December 1991, pp. 39–45.

This article points out that a company can benefit by prematurely making its profitable product obsolete despite a contrary indication from a pure DCF analysis.

APPENDIX E

Books

Quality Function Deployment Three Day Workshop: Implementation Manual, Version 3.4. Dearborn, MI: American Supplier Institute, 1989.

Transactions from the Third Symposium on Quality Function Deployment, June 24–25, 1991, Novi, MI.

Article

Abbie Griffin. "Evaluating QFD's Use in US Firms as a Process for Developing Products." *Journal of Product Innovation Management,* Vol. 9, No. 3, September 1992, pp. 171–187.

APPENDIX F

M. D. Rosenau. *Successful Project Management,* 2nd ed. New York: Van Nostrand Reinhold, 1992, pp. 67–69, 79–82, 96–101, 133–135, 214–216, 272–275.

Index